RF Design Guide

Systems, Circuits, and Equations

For a complete listing of the *Artech House Microwave Library*,
turn to the back of this book.

RF Design Guide

Systems, Circuits, and Equations

Peter Vizmuller

Artech House
Boston • London

Library of Congress Cataloging-in-Publication Data
Vizmuller, Peter, 1954–
    RF design guide : systems, circuits, and equations / Peter Vizmuller.
        p.   cm.
    Includes bibliographical references and index.
    ISBN 0-89006-754-6
    1. Radio circuits–Design and construction–Handbooks, manuals, etc. I. Title
II. Title: Radio
Frequency design guide.
TK6560.V58   1995   94-47118
621.384'1–dc20   CIP

British Library Cataloguing in Publication Data
Vizmuller, Peter
RF Design Guide: Systems, Circuits and Equations
I. Title
621.38412

ISBN: 0-89006-754-6

© 1995 ARTECH HOUSE, INC.
685 Canton Street
Norwood, MA 02062

International Standard Book Number: 0-89006-754-6
Library of Congress Catalog Card Number: 94-47118

10 9 8 7 6 5 4 3 2 1

Contents

Preface

In a perfect world, there would always be enough time to design a system or a circuit. The appropriate books and experts could be consulted, and the particular design approach could be justified to the customer or to upper management by means of clever computer simulations. Engineers and scientists could quantify and resolve design risks and settle for win-win compromises.

Well, you and I know that our world is not like that. Shorter development cycles, uncompromising emphasis on product quality, changing system requirements, and the introduction of new technologies all place demands on the design engineer's time. Lack of sufficient design time can lead to a paradox: we reinvent the same circuits because there is not enough time to write a detailed design summary of a working circuit, something that would really benefit the next project and educate new engineers.

The purpose of this book is to provide a means out of that paradox by summarizing in one reference the information required for a typical radio frequency (RF) communication project. With this book, you should be able to quickly learn about the five factors that affect receiver adjacent channel selectivity. Look up a mixer, understand its operation and design tradeoffs in about 15 minutes. Learn how to measure intercept point. Find out about image noise and its effect on receiver sensitivity. Evaluate an important waveguide formula to quantify the effectiveness of a perforated shield. The information on most topics is presented at several levels of technical detail to allow you access to as much information as your time or interest allows.

Shorter design cycle times, improved quality, and reduced design risks were the main drivers for this book. Constantly changing technology both necessitates and allows for the material contained in this book to be organized in a computer database, where it can be easily updated and disseminated. The Workbook software included with this book takes a first step in that direction by allowing you to quickly look up and evaluate virtually all of the 382 equations described in the book.

Because RF and microwave applications are growing by leaps and bounds, it is inevitable that some important circuits and system concepts could be added to this book in the future. I would therefore appreciate your input on additional important topics, as well as your comments on the ultimate electronic database. I can be reached at the following address: 207 Harding Blvd. W., Richmond Hill, Ontario, Canada L4C 8X6.

I would like to express my appreciation to my colleagues E. Whitehead, W. Sierocinski, and Z. Jagaric, as well as the editors and the reviewers at Artech House, for their many helpful hints and suggestions. My wife, Suzanne, and our children, Catherine, Andrew, and Deanna, deserve special thanks for their patience.

Introduction

The general format of this book is that of a handbook, a miniencyclopedia of RF engineering. Therefore, the topics in Chapters 2, 3, and 4 are arranged in alphabetical order to allow faster and more convenient access to information. The chapters themselves are organized to follow a typical design sequence:

- *What do I need?* Determine the circuit specifications from the system architecture requirements described in Chapter 1.
- *Which circuit do I use?* Consult Chapter 2 for representative circuits.
- *How do I measure it?* Use techniques from Chapter 3 to verify circuit performance.
- *What are the practical and theoretical limits on the behavior of my circuit?* Chapter 4 contains important equations that help you quantify design decisions.
- *How do I quickly evaluate this complicated formula?* Enjoy the convenience of hundreds of preprogrammed and verified equations in the Workbook software.

The sections in Chapter 1 on receiver and transmitter architectures rely heavily on material contained in the rest of the book, and you may want to return to these two sections after becoming familiar with the rest of Chapter 1. These sections relate system requirements to circuit block performance specifications and examine the various tradeoffs involved. All equation variables and their units of measurement are defined following each equation, and all specialized terms and acronyms are summarized in the glossary.

Linear quantities are always shown in italics, while plain text is used for logarithmic quantities, to avoid confusion between such quantities as noise factor and noise figure, linear gain and gain in dB. Operators (e.g., Δ) and units (e.g., Hz) typically do not appear in italics. Several symbols are only used for specific quantities: λ always refers to wavelength, ω to radian frequency, Γ to reflection coefficient.

Personal computers, with their capability for convenient computation, have freed us from approximate nomographs and tables, a trend reflected in this book: I have tried to stay away from approximations and used the exact expressions whenever possible. An example that illustrates the advantages of this approach is in Section 4.18, which contains the generating equations for the Smith Chart; you can generate any portion of the Smith Chart at any desired resolution using those equations and high–end graphics software. Manipulation of s-parameters is another example; the approximation of unilateral transistor operation in order to simplify the mathematics is no longer necessary or justified.

The topics in this book have been selected to be generally useful in real-life radio frequency designs. Such designs are often complex and involve many tradeoffs; this book assumes that the reader is basically familiar with high-frequency electronics, applications of the Smith Chart, complex algebra, and computer simulation. Some topics have been deliberately left out or covered in less detail, because adequate coverage is available elsewhere in the literature. Basic theory of RF circuits [1], transmission line design and analysis [2], filter design [3], synthesizer design [4], electromagnetic [5] and antenna theory [6] are good examples of topics this book is intended to complement.

Several examples of computer simulations are used for illustration purposes throughout the book. In the past, authors were careful to avoid endorsing a particular product or application. For many reasons, the actual applications change all the time, but I think that the idea of computer simulation is here to stay. I would like to endorse computer simulation in general: it is an excellent educational tool that allows engineers to pursue and verify ideas far faster and more conveniently than possible on the bench. Statistical analysis provides information that cannot be easily obtained by bench testing. At the same time, engineering judgment—the ability to weigh contradictory requirements in evaluating system and circuit feasibility—will always require the creative human touch of an experienced engineer.

The Workbook software that accompanies this book is a result of my efforts to verify the hundreds of equations used throughout the book by programming them into a spreadsheet environment. I realized that my collection of spreadsheet calculations should be included with the book, to offer readers convenient numerical evaluation of equations, organized and numbered in the same sequence as the book's table of contents.

REFERENCES

[1] Bowick, C., *RF Circuit Design*, 1st ed. Carmel, IN: Sams Books, 1982.
[2] Wadell, B. C., *Transmission Line Design Handbook*. Norwood, MA: Artech House, 1991, p. 25.
[3] Zverev, A. I., *Handbook of Filter Synthesis*. New York: John Wiley and Sons, 1967.
[4] Rohde, U. L., *Digital PLL Frequency Synthesizers—Theory and Design*. Englewood Cliffs, NJ: Prentice–Hall, Inc., 1983.

[5] Kraus, J. D., and K. R. Carver, *Electromagnetics*, 2d ed. New York: McGraw-Hill, 1973, p. 404.
[6] Johnson, R. C., and H. Jasik, *Antenna Engineering Handbook*, 2d ed. New York: McGraw-Hill, 1984.

CHAPTER 1

System Design and Specifications

1.1 RECEIVER DESIGN

The purpose of a radio frequency (RF) receiver is to process incoming energy into useful information, adding a minimum of distortion. How well a receiver performs its purpose is a function of the system design, its internal circuitry, and its working environment. The acceptable amount and type of introduced distortion vary with the application. Television systems should maintain a signal-to-noise (S/N) ratio greater than 40 dB and are much less tolerant of group delay distortion than analog FM voice systems, for example. Receiver performance specifications in common usage are listed below.

- *Receiver sensitivity* quantifies the ability of a receiver to respond to weak signal levels. The requirement for analog receivers is maximum RF level to ensure a certain demodulated S/N ratio, while digital receivers use maximum bit error rate (BER) at a certain RF level as a measure of performance. The receiver sensitivity specification and its measurement rely on the assumption that thermal noise is the only limiting factor; this assumption may not be valid for systems operating at lower frequencies. In such systems, the input S/N ratio to achieve the required demodulated S/N output is a more useful parameter, as long as the ambient nonthermal noise level can be measured or estimated. Nonthermal and especially manmade noise do not have well-defined properties; the reliability of a receiver link in such systems is best described in statistical terms, for example, "The output S/N will be better than 20 dB for 95% of the time at a certain location." The actual receiver sensitivity number is correlated to the square root of the receiver's bandwidth and ranges from 0.3 μV for narrowband systems to 0.5 mV in television receivers.
- *Receiver selectivity* usually refers to a receiver's ability to reject unwanted signals on adjacent channel frequencies. This specification, which ranges from 70 dB

to 90 dB, is sometimes so difficult to achieve that many systems do not allow simultaneous active adjacent channels in the same geographical area or on the same cable system.

- *Spurious response rejection* is necessary because all superheterodyne receivers have the potential for responding to frequencies other than the desired channel. This tendency needs to be minimized by the proper choice of the IF (intermediate frequency) and by use of RF filters. Spurious response rejection of 70 to 100 dB can be achieved in practical receivers.
- *Intermodulation rejection* measures the receiver's tendency to generate its own on-channel interference from two or more strong off-channel signals. Seventy decibels is easily achievable, while 90 dB is considered exceptional. Good receiver sensitivity and high intermodulation (IM) rejection are usually contradictory requirements, and some amount of compromise may be required between the two.
- *Receiver self-quieting* refers to reduced receiver sensitivity on some channels due to internally generated signals that capture the detector and thus prevent or inhibit the reception of a desired, weak signal.

Other receiver parameters, such as distortion of the demodulated signal, *S/N* ratio for strong signals (also referred to as *Hum & Noise*), frequency stability, cochannel rejection, cross-modulation, radiated emissions, and susceptibility to high RF levels, are also prominent in many regulatory requirements because of their importance in dense carrier environments.

Environmental conditions, including operating temperature range, humidity, weatherproofing, and shock and vibration requirements, can impose severe limitations on the component selection and mechanical packaging.

Production yields, reliability, quality, and easy servicing round off the list of receiver requirements. Many of these requirements have become specialized disciplines on their own; additional information can be found in [1] and [2].

1.1.1 Receiver Architecture

Formulas and equations for designing a particular circuit are useful once we determine that a certain level of performance is required. But how do we determine what level of performance is in fact necessary? The purpose of this section is to examine a typical receiver architecture, explaining the functions of the various circuit blocks. The primary motivation for selecting a certain receiver architecture is the required performance. A garage door opener will likely have different architecture from a high-performance base station receiver, and the design effort for a particular receiver is a very steep function of the required performance specifications.

This section concentrates on high-performance receivers, because more considerations need to be taken into account.

It may be instructive to walk through Figure 1.1 block by block and summarize the important design decisions for each block before tackling the somewhat iterative procedure for determining individual stage specifications.

The antenna *must* be connected to a dc ground either by the structure of filter 1 or by a separate RF choke or high-valued resistor. The antenna is exposed to an uncontrolled environment, and any static charge accumulated on it will translate to very high voltage in internal receiver components due to the low value of capacitors used at RF. (You will recall that $V = Q/C$; the voltage induced by a given amount of charge on a small capacitor is larger than voltage induced on a large-value capacitor. Many pieces of equipment can be seen with a choke soldered to the antenna terminal, evidently done after some field problems were encountered!)

Filter 1 is usually called the preselector and has three basic functions: to limit the bandwidth of spectrum reaching the RF amplifier and mixer to minimize IM distortion; to attenuate receiver spurious responses (image and 1/2 IF are most important); and to suppress local oscillator energy originating in the receiver. Attenuation of direct IF frequency pickup may also be a concern in receivers with high first IF frequency. RF filter 1 may be a highly selective, cavity tuned filter, cascaded with a low-pass filter to attenuate reresonances at odd multiples of the center frequency (a property of all such filters).

RF amplifier noise figure, gain, and intercept point are set by receiver performance requirements. High reverse isolation is important to attenuate local oscillator energy and to isolate filter 1 and filter 2 from each other, so that overall selectivity is not destroyed. Low reverse isolation in the RF amplifier will cause the filters to interact, with guaranteed degradation of RF selectivity at some frequencies.

The function of filter 2 is to attenuate receiver spurious response frequencies, attenuate direct IF frequency pickup, attenuate noise at the image frequency originating in or amplified by the RF amplifier, and suppress second harmonic originating in the RF amplifier, which could potentially degrade mixer second-order intercept point. Depending on its bandwidth, filter 2 can also suppress local oscilla-

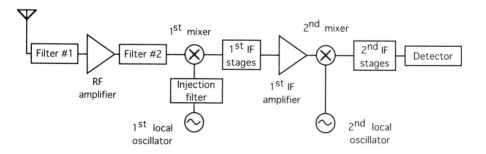

Figure 1.1 Typical dual-conversion receiver.

tor energy leaking back into the antenna. It is important for this filter not to have any return responses at high frequencies, because the mixer has very little rejection for odd harmonics of the receive frequency that may leak through the system. This filter is frequently called the image filter because it typically rejects image noise by about 20 dB. If good receiver sensitivity is not a requirement, the RF amplifier and image filter may not be required.

The first mixer is by definition a nonlinear device, and it usually encounters the highest RF levels present in the receiver. It therefore needs to have a high intercept point. From the wide variety of mixers available (active, passive, unbalanced, singly balanced, double balanced, tuned, broadband), choose the one that best meets all the requirements. In general, passive, double-balanced mixers have the highest intercept point, best noise balance, and highest local oscillator power requirement. Single-device active mixers are the cheapest but have poor intercept points. Table 1.1 summarizes important mixer properties and their effect on receiver performance.

A diplexer network is frequently used at the mixer IF port to optimize intercept point performance, by not allowing any signals, especially local oscillator (LO) harmonics, to be reflected back into the mixer. The diplexer network must be nonreflective up to several times the LO frequency.

The injection filter may be required to attenuate wideband noise around the LO frequency and its harmonics and to attenuate the second harmonic in order not to degrade mixer second-order intercept point (IP2). Since the LO signal is of high amplitude and performs some switching action inside the mixer, the natural tendency would be to ignore incoming harmonics on the LO port, because the mixer internal switching action will generate its own harmonics anyway. This line of reasoning fails for double-balanced mixers, because they are internally balanced against the second harmonic and theoretically do not generate any even harmonics internally. Therefore, the attenuation of externally generated second harmonic of the LO signal is desirable in double-balanced-mixer applications.

Table 1.1
Important Mixer Parameters

Mixer Parameter	*Affected Receiver Specification*
Conversion loss	Receiver sensitivity
Third-order intercept point	Intermodulation distortion
Second-order intercept point	1/2 IF spurious response rejection
Higher-order intercept point	High-order spurious rejection
Noise balance	Receiver sensitivity, AM noise rejection
LO to RF isolation	Conducted LO energy propagating toward antenna
RF to IF isolation	Susceptibility to direct IF frequency pickup

An important property of the first local oscillator is its single-sideband (SSB) phase noise, which often determines the receiver's adjacent channel selectivity performance. The wideband noise, which is typically measured at frequency offsets much greater than the SSB phase noise, affects receiver sensitivity. The LO signal must have low spurious signals; if present, they will cause receiver spurious responses. The first LO synthesizer is the limiting circuit block for frequency change lock time. The LO must oscillate despite temperature and power supply variations. Another requirement in mobile or portable equipment is low susceptibility to microphonics, where external mechanical or acoustic stress could modulate the LO frequency or amplitude.

The first IF (usually crystal) filter protects its following stages from close-in IM signals, provides adjacent channel selectivity, and attenuates the second image. Frequently, the second-image requirement is more stringent than the adjacent channel selectivity requirement and determines the number of poles required to obtain the required second-image selectivity. The equivalent noise bandwidth of the IF chain is an important receiver property, because it determines how much noise reaches the detector, and it determines the modulation bandwidth that can be received. Low group delay of the IF filters is particularly important for digital communication. Group delay compensation in hardware or software can be used to overcome the undesirable effects of group delay distortion, provided the group delay is highly repeatable from unit to unit. The very first IF crystal filter following the mixer may have to be selected for good IM. It may also have to be isolated from the mixer by an impedance inverter network to limit the maximum impedance presented to the mixer on the filter skirts, because high impedance at the mixer IF port will degrade its IM performance. This is especially important for active mixers.

The IF amplifier is usually a high-gain stage. Its intercept point must be high if it directly follows the mixer. If it follows one stage of IF filtering, the intercept point requirements can be relaxed, because the IF filter offers some protection against high-level, off-channel signals.

The steps required for deciding the specifications for individual circuit blocks shown in Figure 1.1 can be summarized as follows. Keep in mind that the whole process is iterative. You may have to rethink a particular strategy if you discover other tradeoffs during the receiver definition. It is assumed that the demodulation process is taking place in an integrated circuit with well-behaved performance characteristics.

1. Allocate approximate gains and losses as needed to meet the required receiver sensitivity specification and IM distortion requirements.
2. Select the first IF frequency.
3. Select the first LO injection side.
4. Investigate the mixer.
5. Based on mixer performance, design the injection filter and select LO technology.

6. Investigate filter topologies.
7. Design the RF amplifier.

A first estimate of the required gains, losses, and intercept points will show if the system requirements for sensitivity and IM rejection are feasible. A computer spreadsheet using formulas in this book and included in the Workbook software can be invaluable in examining different receiver approaches. Subsection 1.1.2, which discusses receiver sensitivity, contains an example of all the required circuit and system parameters required to calculate receiver sensitivity. In general, RF amplifier gains in excess of 20 dB are undesirable because the required gain may not be available from a single device, stability may be compromised, and the resulting mixer intercept point requirement may not be achievable. Filter insertion loss of up to 3 dB is typical. Passive versus active mixer selection may have to be examined at this time as well. Many receivers with active mixers do not require an RF amplifier.

Selection of the IF frequency will dictate the other filter specifications, because the IF frequency determines the location of image and 1/2 IF spurious response frequencies. Crystal or other IF filters come only in certain center frequencies; however, if you can choose among several standard IF frequencies, choose the highest frequency. In addition, examine the high-order spurious responses of the mixer to determine the lowest order of such spurs that fall in-band and make sure the IF frequency is not a harmonic of a digital clock, reference frequency, or any other discrete frequency already present in the vicinity. This also includes external signals.

The selection of the first LO injection side has the following three considerations:

1. High-order spurious responses and self-quieting frequencies may favor one or the other injection side, once the IF frequency has been chosen.
2. Higher-frequency oscillators typically have worse SSB phase noise, but the required voltage-controlled oscillator (VCO) tuning range (in percent) for synthesized sources is less for high-side injection than for low-side injection. The chosen mixer may have a limited frequency of operation, forcing low-side injection.
3. A lower-frequency LO that is multiplied up in frequency may sometimes offer advantages over a high-frequency LO without frequency multiplication.

Mixer performance is probably the single most important determinant of receiver performance. Passive mixers generally have better IM performance but require much higher local oscillator power and do not provide conversion gain. Active mixers require less local oscillator power, and, contrary to expectation, they do not have much better noise figure that passive mixers. A high third-order intercept point (IP3) requirement may compromise the power dissipation specification, especially at high temperature. The mixer noise balance will determine if an injection filter is required. Mixer second-order intercept point will dictate what amount of RF filtering will be required for the 1/2 IF spurious response. Selection of mixer technology will dictate the LO power requirement. The more amplification of the VCO signal that is required, the higher the wideband noise will be, so that high LO powers may require an injection filter to suppress image noise if excellent sensitivity is to be achieved. If the LO is required to tune over a wide frequency range, the injection filter may have to be tunable to track it. Regardless of the mixer noise balance and LO wideband noise performance, using some form of injection filter may still be advantageous. For example, a simple low-pass filter, which cuts the second harmonic from the LO signal, may improve the mixer IP2 and will attenuate noise sidebands around the second harmonic, where the mixer noise balance may not be particularly good.

The LO technology is probably dictated more by the actual application than by anything else in the receiver. If the receiver's frequency is expected to be programmable, a frequency synthesizer is required. Single-frequency receivers can get by with a crystal oscillator. A *LC* discrete inductor-capacitor oscillator circuit can be used for the LO in the ultimate low-cost receivers, as long as the transmitter frequency can change its frequency to compensate for receiver frequency drift or vice versa. One such scheme may have a transmitter that repeatedly transmits data bursts while its carrier frequency changes linearly between two limits. As long as the intended receiver can receive signals somewhere between the same two limits, it will receive the message.

The RF filter requirements will be determined by the chosen IF frequency and first LO injection side. Using high-side injection, all the important reject frequencies will be on the high-frequency side of the passband and vice versa. Therefore, a filter topology that rejects the appropriate signals has to be selected. The simple examples in Figure 1.2 and Figure 1.3 illustrate two practical filter topologies with steeper skirts on the low- (Figure 1.2(a)) and the high-frequency side (Figure 1.2(b)). The usual tradeoff in filter performance is selectivity versus insertion loss. Low insertion loss is more important before the RF amplifier; it can be sacrificed for selectivity if the filter follows the amplifier.

Typically the RF amplifier gain is fine-tuned once the properties of all the other circuits are known. This can be achieved with the RF amplifier much more easily than with the other stages. For example, once a physical size for a cavity tuned filter is chosen, its unloaded Q and therefore insertion loss are basically fixed, and very little can be done to change them later. Similarly with the other RF filters; inductor Q and the number of stages fix the insertion loss. Mixer properties are not easy to change either. In contrast, a wide variety of techniques are available to change an amplifier's gain or intercept point. Thus, the RF amplifier design can be fine-tuned as the last step.

The second IF circuitry, together with the demodulator, is very likely to be contained in an integrated circuit and thus is not under direct control. Things to

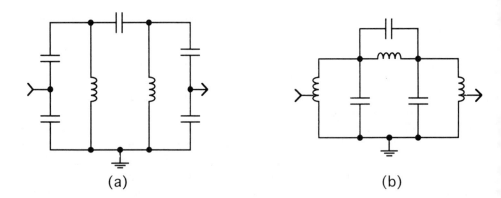

Figure 1.2 Two filter topologies with different skirt slopes.

look for are gain, noise figure, intercept point, amplitude-limiting properties, and integrated circuit (IC) properties such as input impedance, current drain, and bypass requirements. Frequently, the IC will require particular filters and matching networks specified by its manufacturer.

While the second IF circuitry is not critical, and the design requirements are quite loose, there is one important exception: the *S/N* ratio at detector input, required for some particular receiver baseband performance (12 dB SINAD, 20 dB T/N, BER, etc.). This detector property goes under various names, such as cochannel rejection, capture ratio, and rise number. This is a critical parameter for any receiver design and should be as low as possible. It is basically the IF *S/N* ratio at the detector input, required to meet some of the baseband specifications.

Receivers with low power consumption have additional constraints on their operation. Such receivers are always susceptible to overload and IM distortion and are best designed as narrowband as possible. Frequently such receivers are cycled on and off to conserve battery power. When cycling the dc power to a circuit, all the capacitors need to be charged and discharged, introducing delay and ultimately placing a limit on how fast and for how short a time the circuit can be turned on. The most surprising offenders are coupling capacitors in the IF section. Imagine that you are coupling the IF through a 0.1-μF capacitor into an integrated circuit whose input impedance is 5 kΩ. The IC will not have the correct bias voltage at its input pin and will not operate correctly in the approximately 1 ms that the capacitor needs to charge. Lower-value coupling capacitors and precharging schemes can help.

1.1.2 Receiver Sensitivity

Receiver sensitivity is a fundamental property that intimately affects system performance. The following analysis assumes that the receiver's sensitivity is limited by

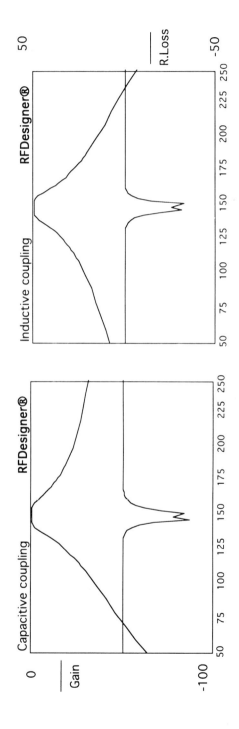

Figure 1.3 Frequency response plots of two bandpass filter topologies.

thermal noise, except as noted otherwise, which really represents the best case. In many cases, ambient noise is higher than thermal, and IM products produced in a dense carrier environment place an effective lower limit on signal amplitude in a workable system.

The main theme carried throughout this section is that device gains and noise figures, image noise, and local oscillator wideband noise need to be treated separately and then combined to produce an overall equivalent input noise factor, F_T, which is then used in (1.2) to calculate the overall receiver sensitivity.

$$F_T = F_{IN} + F'_{IN} + F''_{IN} \tag{1.1}$$

F_T = total equivalent input noise factor (linear)

F_{IN} = total equivalent input noise factor derived from on-channel stage noise figures and gains (linear)

F'_{IN} = total equivalent input noise factor derived from image frequency stage noise figures and gains (linear)

F''_{IN} = total equivalent input noise factor derived from local oscillator wideband noise (linear)

$$e = \sqrt{F_T k\, T\, B\, (R - 1)\, R_G} \quad \text{for receivers limited by thermal noise} \tag{1.2}$$

$$e = \sqrt{k(T_i + T_e)\, B(R - 1)\, R_G} \quad \text{for receivers limited by non-thermal noise} \tag{1.3}$$

e – receiver sensitivity (V)

F_T = total equivalent input noise factor (linear)

k = Boltzmann's constant, 1.38×10^{-23} (J/K)

T = temperature (K), $T(K) = T(°C) + 273.15$

B = equivalent noise bandwidth of system (Hz)

R = required $(S + N)/N$ at detector input (linear)

R_G = system impedance (Ω)

T_i = antenna input temperature (K)

T_e = equivalent receiver noise temperature (K), which is $(F_T - 1)$ 290

The S/N ratio required at the input of a detector depends on the particular measure of receiver sensitivity. For example, 12 dB SINAD (audio measurement) may require 5 dB S/N ratio at detector IF frequency. Detector IF S/N ratios are different for alternate measures of sensitivity, such as BER, quieting, and tone/noise ratios. The receiver sensitivity is then calculated for the desired detector S/N ratio. The quantity $R = (S + N)/N$ is used in the equations because it is easier to measure. The quantities $(R - 1)$ and (S/N) are mathematically identical.

1.1.2.1 Contribution From Stage Gains and Noise Figures

Noise figures and gains of all stages up to but excluding the detector must be known to calculate the equivalent input noise factor. The noise figure of passive stages, which do not contain noise sources other than thermal noise, equals their loss in decibels, or their noise factor equals reciprocal of gain.

$$F_i = \frac{1}{G_i}$$

F_i = stage noise factor (linear)

G_i = stage gain (linear)

If a passive stage is at a higher temperature than the rest of the receiver chain, its noise figure must be adjusted for the temperature difference [3]. Adopt the lowest component temperature as the system noise temperature and assign higher noise factors to all components at a temperature higher than the system temperature. The noise factor for a passive (lossy) device at any temperature is

$$F = 1 + (L - 1)\frac{T}{T_0} \tag{1.4}$$

F = noise factor of lossy device (linear)

L = loss of device = 1/gain (linear)

T = device physical temperature (K)

T_0 = room temperature, defined as 290K

Insertion loss before first gain in the signal path adds its loss to the noise figure of the gain stage decibel for decibel. The gains and noise factors of active stages are not correlated and must be obtained separately. Some passive stages, such as double-balanced diode mixers, can have a noise figure that is slightly higher than their loss.

The stage gains and noise figures are usually expressed in decibels. To convert from decibels to linear units:

$$F_i = 10^{(F_i/10)}$$

$$G_i = 10^{(G_i/10)}$$

F_i = stage noise factor (linear)

G_i = stage power gain (linear)

F_i = stage noise figure (dB)

G_i = stage gain (dB)

Friis formula for calculating cascaded noise figure is used to combine the stage contributions and can be generalized for n stages.

$$F_{IN} = F_1 + \frac{F_2 - 1}{G_1} + \frac{F_3 - 1}{G_1 G_2} + \ldots \tag{1.5}$$

$$F_{IN} = 1 + \sum_{i=1}^{n} \frac{(F_i - 1)}{\prod_{j=0}^{i-1} G_j} \tag{1.6}$$

F_{IN} = equivalent input noise factor (linear)

F_i = stage noise factor (linear)

G_i = stage gain (linear)

$$\prod_{j=0}^{i-1} G_j = \text{total prestage gain (linear)}$$

n = total number of stages up to but excluding detector

i = index in summation of stage terms

j = index in product of prestage gains, $G_0 = 1$

Subtract 1 from the noise factor of each stage, active or passive, and divide by the total prestage gain. Then add 1 to the sum of all such stage contributions to arrive at the equivalent input noise factor contributed by stage gains and noise figures.

1.1.2.2 Contribution From Image Noise

Image noise is simply noise contained at the receiver's image frequency, present at the first mixer's RF port, which adds to the downconverted signal produced by an on-channel signal on a power basis. Therefore, the analysis only needs to be carried out up to the mixer stage that combines the two sidebands into one IF signal. Figure 1.4 shows the various frequency relationships between RF, LO, and image frequencies, as well as their noise contributions. Image noise is downconverted to IF with similar conversion loss as the desired signal.

Image noise can originate from the outside or be the internally amplified thermal noise at the image frequency. For example, the RF amplifier will amplify its own noise as well as any input noise not only near the required frequency but also at the image frequency. If this image noise is filtered out before it reaches the mixer, sensitivity improvement can be achieved. In a well-designed receiver, the image noise coming from the antenna is filtered out and can be neglected, but internal noise generated in the preamplifier cannot be neglected. Image noise can be treated by finding an equivalent on-channel input noise source. We have to be careful to sort out the noise factors and gains that apply to the image frequency and the ones that apply to the on-channel frequency. The reason for the complication is our aim to replace an image noise signal with its on-channel input equivalent. Therefore, the normal analysis carried out at the image frequency has to be normalized by the ratio of the overall image frequency gain to on-channel frequency gain. When a receiver uses high first IF, the gains and noise figures at the image frequency may be significantly different from their on-channel counterparts and have to be measured or calculated at the proper frequencies.

In multiple-conversion receivers (which have more than one IF), additional image noise calculations may need to be carried out, depending on the gain distribution in the receiver chain. In a well-designed receiver with a high-gain first stage, the noise floor at the second or third mixer input should be predominantly input noise amplified by the first stages, and thermal noise amplified by subsequent stages can be neglected. If that is not the case, image noise analysis at the second and third image frequencies may need to be carried out as well.

The conventional cascaded noise figure analysis assumes that all the stages are conjugately matched to each other and that all available noise power from one stage is totally absorbed by the next stage without reflection. This assumption is definitely not true if we deliberately introduce filtering to eliminate image noise. The fundamental difference in the two approaches is that in the matched case, the noise level can never fall below thermal levels, whereas in the mismatched case, image noise can be below thermal. In the matched case, the noise figure of a passive stage is equal to its loss. In the mismatched case, we must allow zero noise figure to be assigned to a passive stage, which provides image attenuation by reflection. If the mixer is looking into a totally reactive termination at its input at the image frequency, it produces no thermal noise and hence can be assigned zero noise figure. The example in Subsection 1.1.2.4 clarifies these concepts.

Image noise contribution to equivalent input noise factor is

$$F'_{IN} = \frac{\prod_{i=1}^{N} G'_j}{\prod_{i=1}^{N} G_j} \left[1 + \sum_{i=1}^{N} \frac{(F'_i - 1)}{\prod_{j=0}^{i-1} G'_j} \right] \qquad (1.7)$$

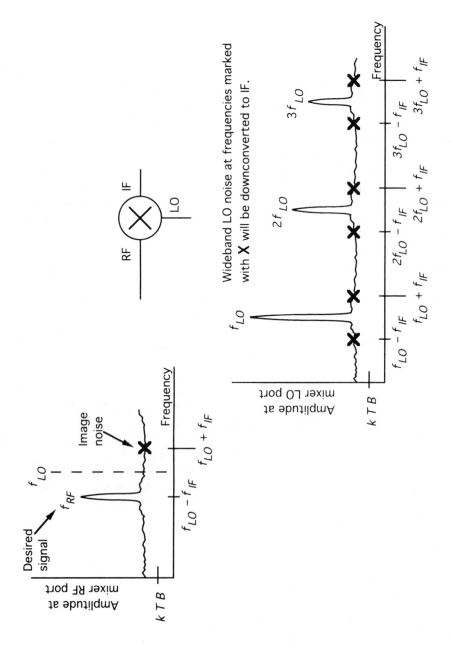

Figure 1.4 Image noise and wideband LO noise contributors to overall noise figure degradation.

F'_{IN} = contribution to overall input noise factor by image noise (linear)

F'_i = stage noise factor at image frequency (linear)

F'_i = 1 for the stage that provides image attenuation by reflection (ideally the stage immediately preceding the mixer)

G'_j = stage gain at image frequency (linear), $G'_0 = 1$

G_j = stage gain on-channel (linear), $G'_0 = 1$

N = number of stages up to but excluding the mixer

The implicit assumption in (1.7) is that the mixer conversion loss is the same at the desired frequency as at the image frequency. High-performance mixers are often implemented with attenuators at all three ports to optimize third-order-intercept performance by not allowing harmonics to be reflected back into the mixer for remixing, as shown in Figure 1.5.

It should be recognized that the attenuator at the mixer RF port partially negates the benefits of the image filter, because it reintroduces thermal noise at the image frequency, which normally would be eliminated by the image filter. Evaluating (1.7), we can still assign zero noise figure to the image filter, but the noise figure of the RF port attenuator will be equal to its attenuation at the image frequency.

1.1.2.3 Contribution From Wideband LO Noise

The wideband noise of the local oscillator is another parameter that can elevate IF noise level, thus degrading the overall noise figure. The mechanism of this noise generation is that wideband noise separated from the LO frequency by $\pm f_{IF}$ spacing will mix to produce noise at the IF frequency. This noise conversion process is related to, but not the same as, LO to RF isolation. Noise at frequencies of $\pm f_{IF}$ spacing from the LO harmonics also contributes and may be dominant in some cases, as will be discussed in Section 3.8. Figure 1.4 shows the frequency relationships

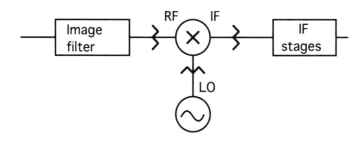

Figure 1.5 Attenuators at mixer ports to improve IP3.

among the noise sidebands and LO harmonics. In contrast to image noise, wideband LO noise is downconverted to IF with a much higher conversion loss than the desired signal.

Wideband noise from all sources adds on a power basis at the mixer IF output. The wideband noise sidebands are often not symmetrical and also not flat with frequency. The noise level in decibels below carrier per hertz has to be measured separately at $nf_{LO} + f_{IF}$ and $nf_{LO} - f_{IF}$ for each n. This conversion of noise sidebands into IF noise is called mixer noise balance; it is really conversion loss for the relevant sideband from the LO port to the IF port.

There may be a bandpass filter between the LO and the mixer; in that case, its loss at the appropriate noise sideband needs to be taken into account.

$$F''_{IN} = \sum_{s=1}^{M} \frac{10^{[(P_{LO}+W_s-L_s-M_s)/10]}}{1000 \ k \ T_0 \prod_{j=1}^{N} G_j} \tag{1.8}$$

F''_{IN} = contribution to overall noise factor by wideband LO AM noise (linear)

P_{LO} = local oscillator power (dBm)

W_s = wideband noise level of sideband s (dBc/Hz)

L_s = loss of injection filter at frequency of sideband s (dB)

M_s = mixer noise balance for sideband s (dB)

k = Boltzmann's constant, 1.380×10^{-23} (J/K)

T_0 = 290K

G_j = gain of stage j (linear)

s = index for summation of noise powers at all sidebands of interest

M = number of sidebands taken into account

j = index to calculate gain up to and including the mixer

N = number of stages up to and including the mixer

If there is no injection filter, L_s = 0 dB. The summation is for all the sidebands that are of interest, mainly, $f_{LO} + f_{IF}$, $f_{LO} - f_{IF}$, $2f_{LO} + f_{IF}$, $2f_{LO} - f_{IF}$, $3f_{LO} + f_{IF}$, $3f_{LO} - f_{IF}$, and so on. Each of these sideband frequencies has its corresponding wideband LO noise contribution, loss of injection filter, and mixer noise balance. The factor of 1,000 in the denominator comes from converting decibels relative to 1 milliwatt to watts, and T_0 comes from converting equivalent noise power per hertz into noise factor.

1.1.2.4 Example of Sensitivity Calculation

The example in Figure 1.6 illustrates application of formulas from the preceding subsections in calculating receiver sensitivity. Assume that we have a dual-conversion receiver with the topology shown in Figure 1.6.

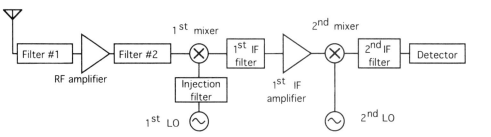

Figure 1.6 Typical dual–conversion receiver.

The stage properties are listed in Table 1.2.

We will proceed to calculate the receiver sensitivity at antenna input by obtaining the equivalent input noise factor due to each of three contributors:

1. On-channel gains and noise figures (F_{IN});
2. Contribution from image noise (F'_{IN});
3. Contribution from local oscillator wideband noise (F''_{IN}).

Using (1.6), we can tabulate the relative on-channel contribution of each stage, as shown in Table 1.3.

Image noise is thermal noise coming through the antenna plus amplified amplifier noise at the image frequency. For the purpose of our analysis, we will assume that the stage gains and the noise figures are the same at the image frequency as on-channel for all stages, except for filter 2, which has a 10-dB loss at the image frequency. We will also neglect any contribution from the second image in the second IF.

Filter 2, which provides image attenuation by reflection, violates the original assumption of a conjugately matched system, where all available noise power from one stage is totally absorbed by the next stage. Therefore, the noise figure of a passive stage is not necessarily equal to its loss in a mismatched system. We will use a noise figure of 0 dB for filter 2 at the image frequency, to allow its output to go below thermal level. Using (1.7), we need to repeat our analysis up to the first mixer stage at the image frequency. The stage properties at the image frequency are listed in Tables 1.4a and b.

The ratio of gain products in (1.7) reduces to the ratio of gains of filter 2, since the gains of all the other stages are the same at the image and on-channel frequencies. Equation (1.7) can now be evaluated:

$$F'_{IN} = 0.1/0.631(1 + 2.98) = 0.63 \tag{1.9}$$

LO wideband noise has six components, thus, $m = 6$ in (1.8). Using the notation of (1.8), P = 23.5 dBm, W_s = −165 dBc/Hz:

Table 1.2
Stage Properties of a Typical Dual-Conversion Receiver

Stage	Gain (dB)	Noise Figure (dB)	Noise Figure (Linear)
Filter 1	−2.5	2.5	1.778
RF amplifier	12.0	3.5	2.239
Filter 2	−2.0	2.0	1.585
First mixer	−8.0	8.3	6.761
First IF filter	−1.5	1.5	1.413
IF amplifier	20.0	4.0	2.512
Second IF filter	−4.0	4.0	2.512
Second mixer	12.0	12.0	15.849
Detector		15.0	31.623

Filter 2 image attenuation	10.0 dB
Equivalent noise bandwidth	12.0 kHz
First LO wideband noise	−165.0 dBc/Hz (flat with frequency)
First LO power	23.5 dBm
Injection filter attenuation	0.0 dB at $f_{LO} \pm f_{IF}$ offset
	10.0 dB at $2\,f_{LO} \pm f_{IF}$
	20.0 dB at $3\,f_{LO} \pm f_{IF}$
Mixer noise balance	30.0 dB at $f_{LO} \pm f_{IF}$
	25.0 dB at $2\,f_{LO} \pm f_{IF}$
	20.0 dB at $3\,f_{LO} \pm f_{IF}$
Required S/N at detector output	6.0 dB (3.981 linear)

Table 1.3
Relative On-Channel Contribution of Each Stage

Stage	Prestage Gain (dB)	Prestage Gain (Linear)	Noise Term
Filter 1	0.0	1.000	0.778
RF amplifier	−2.5	0.562	2.204
Filter 2	9.5	8.913	0.066
First mixer	7.5	5.623	1.025
First IF filter	−0.5	0.891	0.464
IF amplifier	−2.0	0.631	2.396
Second IF filter	18.0	63.096	0.024
Second mixer	14.0	25.119	0.591
Detector	26.0	398.11	0.077

Sum of on-channel noise terms = 7.625.
F_{IN} = 1 + sum of noise terms in (1.6) = 8.625.

Table 1.4a
Stage Properties at Image Frequency

Stage	Gain (dB)	Noise Figure (dB)	Noise Figure (Linear)
Filter 1	−2.5	2.5	1.778
RF amplifier	12.0	3.5	2.239
Filter 2	−10.0	0.0	1.0

Table 1.4b
Image Noise Contributors

Stage	Prestage Gain (dB)	Prestage Gain (Linear)	Noise Term
Filter 1	0.0	1.000	0.778
RF amplifier	−2.5	0.562	2.204
Filter 2	9.5	8.913	0.0
Sum of image noise terms			2.98

1. Noise at $f_{LO} + f_{IF}$: $L_s = 0$ dB, $M_s = 30$ dB.
2. Noise at $f_{LO} - f_{IF}$: $L_s = 0$ dB, $M_s = 30$ dB.
3. Noise at $2 f_{LO} + f_{IF}$: $L_s = 10$ dB, $M_s = 25$ dB.
4. Noise at $2 f_{LO} - f_{IF}$: $L_s = 10$ dB, $M_s = 25$ dB.
5. Noise at $3 f_{LO} + f_{IF}$: $L_s = 20$ dB, $M_s = 20$ dB.
6. Noise at $3 f_{LO} - f_{IF}$: $L_s = 20$ dB, $M_s = 20$ dB.

The gain up to and including the mixer is 0.891 (linear), then from (1.8):

$$
\begin{aligned}
F''_{IN} = {} & 10^{(23.5-165-0-30)/10}/(1{,}000\ k\ T_0\ 0.891) + \\
& 10^{(23.5-165-0-30)/10}/(1{,}000\ k\ T_0\ 0.891) + \\
& 10^{(23.5-165-10-25)/10}/(1{,}000\ k\ T_0\ 0.891) + \\
& 10^{(23.5-165-10-25)/10}/(1{,}000\ k\ T_0\ 0.891) + \\
& 10^{(23.5-165-20-20)/10}/(1{,}000\ k\ T_0\ 0.891) + \\
& 10^{(23.5-165-20-20)/10}/(1{,}000\ k\ T_0\ 0.891) + \\
= {} & 1.984 + 1.984 + 0.628 + 0.628 + 0.198 + 0.198 \\
= {} & 5.62
\end{aligned}
$$

We now have all the information for calculating the total input equivalent noise factor from (1.1):

$$F_T = F_{IN} + F'_{IN} + F''_{IN} = 8.625 + 0.63 + 5.62 = 14.87 \qquad (1.10)$$

Receiver sensitivity can be calculated from (1.2) or (1.11):

$$e = \sqrt{F_T k \; T \; B \left(\frac{S}{N}\right)_0 R_G} \qquad (1.11)$$

$$e = \sqrt{14.87 \times k \; T \times 12000 \times 3.981 \times 50}$$

$$= 0.38 \; \mu V$$

The calculation of (1.10) shows that the LO wideband noise contributes significantly to sensitivity degradation, while contribution from image noise may be neglected.

The series of complex calculations leading to the overall receiver sensitivity number can be conveniently performed on a computer spreadsheet, which can investigate many "what-if?" scenarios.

1.1.3 Receiver Selectivity

Receiver selectivity is a parameter that quantifies the tendency of a receiver to respond to channels adjacent to the desired reception channel. Because international regulations are gradually moving to narrower channel spacings, receiver selectivity assumes greater importance, since it frequently limits system performance and places restrictions on frequency allocation and system utilization.

$$\text{Selectivity} = -CR - 10 \; \log[10^{(-\text{IFsel}/10)} + 10^{(-\text{Spurs}/10)} + BW \times 10^{(\text{SBN}/10)}] \qquad (1.12)$$

Selectivity = amount of adjacent channel selectivity relative to nominal receiver sensitivity (dB)

CR = capture ratio, or cochannel rejection (dB)

IFsel = IF filter rejection at the adjacent channel (dB)

Spurs = LO spurious signals present in the IF bandwidth at a frequency offset equal to the channel spacing (dBc)

BW = IF noise bandwidth (Hz)

SBN = SSB phase noise of LO at a frequency offset equal to the channel spacing Δ (dBc/Hz)

Δ = adjacent channel frequency offset, channel spacing (see Figure 1.7)

For example, assume that the LO spectral purity is as shown in Figure 1.7.

$$\text{Spurs} = 90 \text{ dBc}$$
$$\text{SSB phase noise} = -130 \text{ dBc/Hz}$$
$$BW = 12{,}000 \text{ Hz}$$
$$\text{IFsel} = 100 \text{ dB}$$
$$\text{CR} = 5 \text{ dB}$$

Then, from (1.12), the selectivity would be 81.38 dB.

Selectivity is a property determined by five stage properties. Each of these properties—SSB phase noise, synthesizer spurs, IF selectivity, IF bandwidth, and cochannel rejection—will have a statistical distribution of values. An easy way to ascertain the impact of these statistical variations on overall selectivity is to program the selectivity equation into a system simulator, such as Extend™ [4]. See Section 4.20 for more information on statistical evaluation of system properties.

Receiver adjacent channel selectivity defined by (1.12) assumes that the interfering adjacent channel signal is a perfect, unmodulated sine wave and is thus a purely receiver property. If this assumption is not valid, and the interfering source contains spectral components on the adjacent channel, then the receiver's apparent adjacent channel selectivity performance will be mostly independent of (1.12) and determined by the interferer's spectral characteristics. This situation commonly arises in practice when the SSB phase noise of the interfering signal is particularly poor or in narrowband systems when the interferer's modulating signal contains

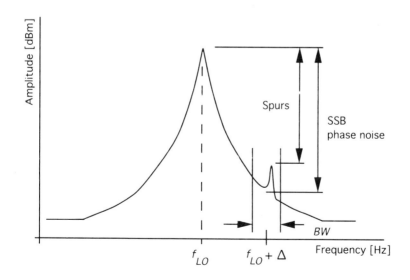

Figure 1.7 Spectral purity of LO signal.

spectral components (splatter) extending into the adjacent channel. This topic is briefly investigated in Section 1.4. A different measure of performance, adjacent channel protection ratio (ACPR), is sometimes used for such systems; ACPR takes into account the interfering signal's spectral characteristics.

1.1.4 Receiver Spurious Responses

Receiver spurious responses are frequencies that are different from the desired receive frequency, yet that still produce demodulated output in the receiver, if encountered at a sufficiently high level. This situation is obviously undesirable and particularly troublesome in receivers capable of tuning over a wide frequency range because the RF filters need to be wide to accommodate the wide frequency coverage.

Most receiver spurious responses are actually mixer spurious responses, which may or may not be further attenuated by RF selectivity in the preceding stages. Most receiver spurious responses result from harmonic mixing of the RF and LO signals. Any RF frequency that satisfies the following relationship is a potential receiver spurious response:

$$\pm m\, f_{RF} \pm n\, f_{LO} = \pm f_{IF}$$

f_{RF} = any incoming frequency into the mixer RF port

f_{LO} = local oscillator frequency

f_{IF} = desired IF frequency

m = integer multiplier of RF frequency

n = integer multiplier of LO frequency

Solving for f_{RF}, each (m,n) pair results in two possible spurious frequencies (all numbers are positive):

$$f_{RF1} = \frac{n\, f_{LO} - f_{IF}}{m} \tag{1.13}$$

$$f_{RF2} = \frac{n\, f_{LO} + f_{IF}}{m} \tag{1.14}$$

f_{RF1} = one possible (m, n) spurious response

f_{RF2} = another possible (m, n) spurious response

f_{LO} = local oscillator frequency

f_{IF} = desired IF frequency

m = positive integer multiplier of RF frequency

n = positive integer multiplier of LO frequency

The following spurious responses are the most common, with (m, n) indices to explain their origin, three of which are graphically illustrated in Figure 1.8.

1. Image: $(-1,1)$ for low-side injection, $(1,-1)$ for high-side injection.
2. Half-IF: $(2,-2)$ for low-side injection, $(-2,2)$ for high-side injection. Amount of rejection can be predicted from the mixer IP2.
3. IF: straight IF frequency pickup.
4. High-order spurs result from combinations of m and n, which result in spurious responses so close to the desired receive frequency that they cannot be filtered out, and mixer performance determines receiver performance. Low-IF frequency receivers will be susceptible to high-order spurious responses where m and n differ from each other by 1. Higher-IF frequency receivers must be examined for susceptibility to spurious responses where the m and n values are not constrained.
5. A whole family of spurious responses of type $(1,n)$ is $n \times$ LO spurs, which can be troublesome if the RF front end has return responses or reresonances, as is often the case with cavity filters [5]. Such filters typically lose selectivity near odd harmonics of the receive frequency, allowing the $n \times$ LO spurs easy access to the mixer.
6. Second image in dual conversion receivers is frequently overlooked, but it often determines how many crystals are required in the first IF to suppress it. In high-performance receivers, the amount of required first IF selectivity may be determined by the second-image rejection requirement, rather than the amount of adjacent channel selectivity. If the second IF is 450 kHz, the second-image spurious response will be $f_{RF} \pm 900$ kHz; the sign is determined by both first and second injection sides.
7. Spurious signals present on the LO signal will cause receiver spurious responses. If the LO is synthesized, it will have sidebands at a frequency offset

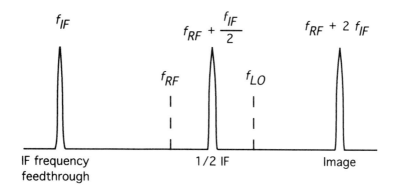

Figure 1.8 Common receiver spurious responses with high–side injection.

equal to the reference frequency, and the receiver will respond to spurious signals offset from the desired receive frequency by the synthesizer reference frequency (both sides) and its harmonics.

8. In full-duplex radios (transmitter and receiver on all the time) the transmitted signal can assume the role of an LO; therefore, the receiver may respond to frequencies separated by the IF frequency from the transmitted frequency. Frequencies used in full-duplex systems must be carefully assigned to make sure the transmitted signal does not fall on any of the normal receiver spurious responses. Otherwise, serious receiver desensitization will occur. The diagram in Figure 1.9 shows two additional spurs in full-duplex radios, duplex image and half duplex:

$$\text{Duplex image} = f_{TX} - \Delta f$$

$$\text{Half-duplex} = f_{TX} + \frac{\Delta f}{2}$$

Δf = frequency difference between receive and transmit frequencies

$\quad = f_{RX} - f_{TX}$

f_{RX} = receive frequency

f_{TX} = transmit frequency

The measurement of receiver spurious responses can be adversely affected by signal generator wideband noise [6]; in case only one generator is used for the

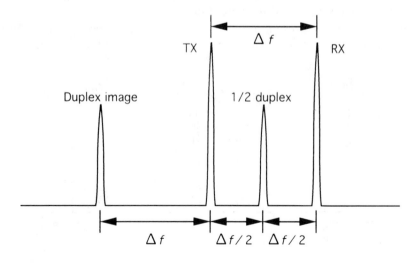

Figure 1.9 Two most common spurious responses in a full–duplex transceiver.

measurement, the receiver will appear to be better than it actually is, while using two generators (one on-channel and the other one interfering) will make the receiver performance appear worse, if signal generator noise is present. When the spurious response of high-performance receivers is being measured, a notch filter tuned to the receiver frequency should be used on the output of the interfering generator to keep the generator's wideband noise from desensitizing the receiver even in the absence of a spurious response signal.

1.1.5 Self-Quieting

Self-quieting is the inability of a receiver to receive a weak external signal, because the detector has already been captured by some internally generated signal. In FM receivers, this phenomenon is manifested as reduced amplitude of squelch noise on a particular channel; the receiver noise is "quieter" there, hence the name.

The most common cause of internally generated interference in dual conversion receivers is harmonics of the second LO. Consider the example shown in Figure 1.10.

Assume the following relationships:

Second IF = 455 kHz

First IF = 21.4 MHz

Second LO = 20.945 MHz (low-side injection)

First LO = 146.16 MHz (low-side injection)

Such a receiver is set to receive an incoming frequency of $146.16 + 21.4 = 167.56$ MHz, but because this frequency is also the eighth harmonic of the second LO ($20.945 \times 8 = 167.56$), the likelihood of internal interference is very high, and the receiver may not be able to pick up a weak external signal at 167.56 MHz.

An example will serve to illustrate the potential impact of such problems. Assume that the receiver sensitivity is −120 dBm and that the best isolation between the second LO and antenna input that can be obtained on our circuit board is 60

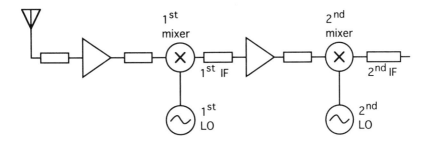

Figure 1.10 Simplified dual-conversion receiver block diagram.

dB. Then, with second-LO power at +10 dBm, we are looking for the eighth harmonic coming out of the second-LO circuit to be about 76 dB down (10 + 120 − 60 + cochannel rejection ratio) in order not to cause interference. Such an amount of attenuation may not be realizable by a filter implemented on a circuit board.

This example also has the potential for another self-quieting response. The seventh harmonic of the second LO is 146.615 MHz, which is exactly 455 kHz above the first LO. In other words, the seventh harmonic of the second LO beating with the first LO fundamental will produce interference at 455 kHz, which is the second IF.

The frequency relationships for generating self-quieting frequencies can often be mind-boggling, especially if we add some microprocessor clocks to the picture. Assume we have a microprocessor running at 1.365 MHz near the same receiver. Then there is a theoretical possibility of generating 21.4 MHz, which is the first IF, by beating together the second harmonic of the first LO, the thirteenth harmonic of the second LO, and the microprocessor clock!

The examples just considered show three of the four possible interference mechanisms: Interference is generated at the on-channel frequency, first IF frequency, second IF frequency, or a receiver spurious response frequency such as the first image.

A good start in the hunt for the self-quieting mechanism is to change the second injection from low side to high side. If the self-quieting disappears, then the second LO is implicated in the problem. Broadband receivers, which are particularly susceptible to self-quieting, deliberately include the capability to use either low- or high-side injection in the second mixer specifically to avoid self-quieting due to harmonics of the second-LO signal. Another troubleshooting aid is to shift the first- or second-LO frequencies in fine increments to observe the effect of the shift. If the IF filter bandwidth is 12 kHz (i.e., ±6 kHz) wide, and you have to shift the first LO by only ±3 kHz for the self-quieting to disappear, then a safe assumption may be that the second harmonic of the first LO is part of the picture.

A computer spreadsheet listing the possible frequencies, their harmonics, and likely mixing products can be invaluable in tracking down self-quieting frequencies.

Microprocessor clocks running at integer subharmonics of either the first or the second IF frequencies should definitely be avoided. Shielding, proper circuit layout, adequate isolation between first and second LOs, and high first IF frequency all help in minimizing self-quieting responses.

1.1.6 Receiver Intercept Point

Intercept point is a measure of circuit or system linearity that allows us to calculate distortion or IM product levels from the incoming signal amplitudes. Input intercept point has been used throughout this book, which does not imply that the output intercept point is not equally valid; the output intercept point equals the input intercept point plus device gain. The input intercept point represents a fictitious input amplitude at which the desired signal components and the undesired components are equal in amplitude, as illustrated in Figure 1.11. The order of the intercept point refers to how fast the amplitudes of the distortion products increase with an increase in input level. For example, for third-order intercept point (IP3) the IM products will increase in amplitude by 3 dB when the input signal is raised by 1 dB.

When a system is analyzed for its overall equivalent intercept point resulting from the combination of its constituent device intercept points, the most frequent assumption is that the distortion products of the various stages are uncorrelated and add on a power basis. If it can be shown that the distortion products in a given connection of devices are correlated (as might happen when a signal is split, processed, and recombined), the phases of the IM signals have to be noted and the signals combined on a voltage basis.

1.1.6.1 Second-Order Intercept Point

Second-order intercept point (IP2) is used to predict mixer performance with respect to a particularly troublesome spurious response, called the half-IF (1/2 IF)

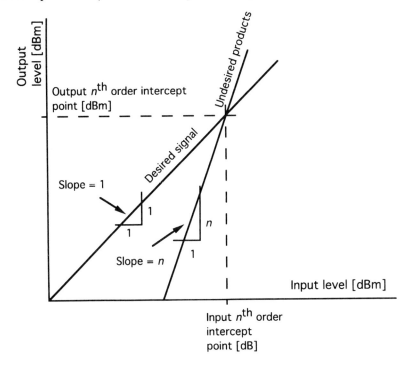

Figure 1.11 Intercept point definition.

spurious response. This receiver spurious response is separated from the desired channel by one-half the IF frequency. The mechanism for 1/2 IF generation is $2f_{RF} \pm 2f_{LO}$, where both harmonics are internally generated, not fed in from the outside.

$$1/2 \text{ IF rejection} = \frac{1}{2}(\text{IP2} - S - CR) \qquad (1.15)$$

IP2 = equivalent input second-order intercept point at receiver input (dBm)

S = receiver sensitivity (dBm)

CR = capture ratio, or cochannel rejection (dB)

For example, say a certain mixer has IP2 = +50 dBm, is used as the first stage of an FM receiver whose sensitivity is −115 dBm, and has cochannel rejection of 5 dB. The 1/2 IF spurious rejection of this receiver would be 1/2(50 + 115) − 5 = 80 dB. Let us assume that this receiver receives 150 MHz and the IF frequency is 21.4 MHz. The 1/2 IF spurious response for the receiver would occur at 160.7 MHz with high-side injection and at 139.3 MHz with low-side injection (150 ± 21.4/2). RF selectivity can be added in front of the mixer to provide, decibel for decibel, more spurious rejection.

An important but frequently overlooked effect is the degradation of spurious rejection when an RF amplifier is placed ahead of the mixer. The origin of this degradation can be intuitively appreciated by considering the preceding example with a 10-dB gain amplifier ahead of the mixer. Such an amplifier would boost all incoming levels by 10 dB, so now the same receiver would respond to a 10-dB lower spurious amplitude; at the same time, however, the sensitivity would not improve by 10 dB, due to the noise figure of the amplifier. Therefore, the amplitude difference between the spurious response and the sensitivity level would be less than without the amplifier.

Mathematically expressed, the RF amplifier reduces the IP2 by the amount of its gain, to +40 dBm, while the sensitivity does not improve by the amount of gain and may only improve to −118 dBm. Using the same formula,

1/2 IF rejection with amplifier = (IP2 − S − CR)/2 = (40 + 118 − 5)/2 = 76.5 dB

The amount in dB by which the RF amplifier degrades spurious rejection can be related to a quantity called the *takeover gain*. The spurious reduction will be equal to (takeover gain in dB)/n, where $n > 1$ is the order of the spurious response ($n = 2$ for 1/2 IF).

$$\text{Takeover gain} = 1 + \frac{F_1 G_1 - 1}{F_2} \text{ (linear)}$$

F_1 = noise factor of RF amplifier (linear)

G_1 = gain of RF amplifier (linear)

F_2 = noise factor of mixer including following stages (linear)

Once the IP2s of all the appropriate circuits are available, the overall system input intercept point can be calculated: Transfer all device IP2s to the system input, subtracting on-channel gains, adding on-channel losses dB for dB, and adding twice the 1/2 IF frequency selectivity (in dB) of the preceding stages for each device. The reason for adding *twice* the 1/2 IF attenuation is the 2:1 slope of second-harmonic-generation curve; if the signal causing the distortion is attenuated, while the on-channel signal is not, the equivalent intercept point improves by twice the amount of 1/2 IF attenuation in dB.

Under normal circumstances, this analysis only needs to include the mixer intercept point, because it is the stage that generates the 1/2 IF spurious response. This assumption would be violated if there were no selectivity at the second harmonic of the RF signal between RF amplifier and mixer. The example in Figure 1.12 illustrates calculation of system IP2, assuming that the mixer is the only nonlinear stage contributing to second-harmonic generation. Note that filter 1/2 IF selectivity refers to attenuation at 1/2 IF relative to the passband; it is not the absolute loss at the 1/2 IF frequency.

The resulting equivalent system input IP2 of +83 dBm would result in about 100 dB of 1/2 IF rejection.

1.1.6.2 Third-Order Intercept Point

Third-order intercept point (IP3) is another important measure of system linearity. It is the theoretical point at which the desired signal and third-order distortion products are equal in amplitude. IP3 determines the amount of IM distortion produced in the receiver itself when subjected to high-level interference.

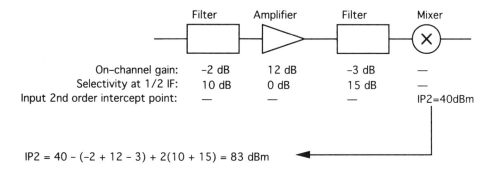

Figure 1.12 Example of system IP2 calculation.

Use the following procedure to calculate equivalent system input intercept point.

1. Draw block diagram of system with associated gains and IP3s.
2. Transfer all intercept points to system input, subtracting gains and adding losses decibel for decibel.
3. Convert input intercept points to powers (dBm to milliwatts).
4. Assuming all intercept points are independent and uncorrelated, add powers "in parallel":

$$IP_{INPUT} = \frac{1}{\dfrac{1}{IP_1} + \dfrac{1}{IP_2} + \ldots + \dfrac{1}{IP_N}} \ [mW] \tag{1.16}$$

5. Convert IP_{INPUT} to dBm (milliwatts to dBm).

$$IP3_{INPUT} = 10 \log\left(\frac{1}{\dfrac{1}{IP_1} + \dfrac{1}{IP_2} + \ldots + \dfrac{1}{IP_N}}\right) \tag{1.17}$$

$IP3_{INPUT}$ = equivalent system input intercept point (dBm)

IP_1 = IP3 of first stage transferred to input (mW)

IP_N = IP3 of last stage transferred to input (mW)

The procedure is best illustrated by the example in Figure 1.13.

$$IP3_{INPUT} = 10 \log\left(\frac{1}{\dfrac{1}{\infty} + \dfrac{1}{15.48} + \dfrac{1}{\infty} + \dfrac{1}{19.95} + \dfrac{1}{100}}\right) = 10 \log(8.02) = 9.04 \text{ dBm}$$

The equivalent system intercept point is 9.04 dBm; the amplifier is the dominant contributor because it has the lowest input equivalent intercept point.

This analysis assumes that there is no RF selectivity for the frequencies causing IM distortion. This situation is, then, slightly different from that in Subsection 1.1.6.1, which deals with IP2s, where in most systems there is usually some RF selectivity for the signals causing second-order distortion. In case there is RF attenuation of the IM signals and no attenuation of the on-channel signal, the IP3 improves by three times the RF attenuation in dB, assuming both IM signals are attenuated by the same amount.

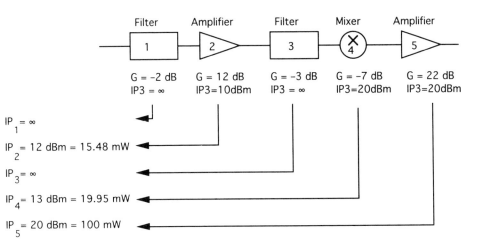

Figure 1.13 Example of system IP3 calculation.

The 1-dB compression point (the level at which the gain drops by 1 dB due to device saturation) is about 11 dB below the IP3 for amplifiers and 15 dB below the IP3 for mixers. The output power of an amplifier at the 1-dB compression point will therefore be about 11 dB below the *output* IP3.

Intermodulation distortion is a property of all systems that exhibit a nonlinear transfer function, such as amplitude compression at sufficiently high levels. Third-order IM distortion is most often produced when two signals, separated by Δf and $2\Delta f$ from the desired carrier beat together and produce on-channel interference. Higher-order IM is also possible.

Intermodulation rejection, which is the difference in decibels between sensitivity and input signal level sufficient to produce interference, can be calculated from the intercept point and receiver sensitivity by use of (1.18).

$$IM = \frac{1}{3}(2 \ IP3 - 2 \ S - CR) \tag{1.18}$$

IM = intermodulation rejection (dB)

IP3 = equivalent input third order intercept point (dBm)

S = receiver sensitivity (dBm)

CR = capture ratio, or cochannel rejection (dB)

In a system with more than two carriers, the number of IM products generated by the $2f_i - f_j$ relationship will be equal to $n \ (n - 1)$, where n is the number of carriers. These IM products will fall near the band of interest; additional $n \ (n - 1)$ products of type $2f_i + f_j$ will be generated near the third harmonic.

The number of triple beats generated by the $f_i + f_j - f_k$ mechanism resulting from n carriers is $n(n-1)(n-2)/2$. Section 1.3 contains additional information on IM resulting from more than three carriers, using a cable communication example.

1.1.6.3 n^{th}-Order Intercept Point

The intercept point concept allows us to predict the distortion products at any input level, provided the distortion products are known for one particular input level.

$$IPn = A + \frac{\Delta}{n-1} \qquad (1.19)$$

IPn = n^{th} order input intercept point (dBm)
A = input signal level (dBm)
Δ = difference between desired signal and undesired distortion (dB)
n = order of distortion

For example, say a certain (4,2) high-order spurious response is 80 dB down when the input level is −20 dBm. (See Subsection 1.1.4 to interpret notation.) Therefore, the fourth-order intercept point is −20 + 80/3 = 6.67 dBm. The (4,2) distortion product for an input level of −30 dBm will then be Δ = (IP4 − A)(4 − 1) = (6.67 + 30) 3 = 110 dB down.

You may have noticed that the order of the distortion product is determined only by the RF multiplier (i.e., 4) and not by the LO multiplier. The reason is that we are concerned only with variations of the RF signal; the mixer LO power is constant in most designs. In general, the distortion products will be strongly correlated to the LO power. Variation of intercept point with LO power can be predicted if we know the mixer diode or transistor properties, but for fixed LO power, this analysis is applicable.

1.2 TRANSMITTER DESIGN

The function of a transmitter is to amplify an RF carrier modulated with the desired signal, adding a minimum of distortion to the encoded information. Specifications that quantify transmitter performance are described next.

Power output is a fundamental communication system parameter whose definition depends on the modulation method used. PM and FM systems use root mean square (rms) power, while AM systems use peak envelope power. Continuous or intermittent duty rating will affect the type of thermal management required to control internal temperatures.

Transmitter turn-on time is important for digital communication systems and must be short in order not to limit system throughput, but not so short as to cause transient energy on adjacent channels. The amount of energy generated by rapid transmitter turn-on and turn-off of the transmitter power can be estimated by applying Fourier Transform theory to such signals. Transmitter power ramp up longer than 6 ms is usually sufficient to minimize emissions at undesired frequencies. Transmitter turn-on time is usually defined as the time for power output to reach 90% of its rated output, but if there is significant amount of load pull, turn-on time refers to the time required for the frequency to settle to some fraction of the channel spacing used.

Load pull refers to the tendency of the transmitter's frequency to shift as a result of changing impedances either at the antenna or internal to the transmitter as the power level builds up.

Spurious outputs are most frequently harmonically related to the main carrier, but other spurious emissions are also common in transmitters whose signal is generated by frequency synthesizers or by a mixing process. Spurious outputs are usually 70 to 90 dB below main carrier; some regulatory requirements specify an absolute maximum power allowed at any frequency other than the carrier.

S/N, Hum & Noise describe the signal-to-noise ratio of the transmitted signal. The *S/N* is usually limited by wideband noise far in excess of thermal produced in the transmitter circuits. Power supply hum, noise pickup, and nonlinearities in the modulation circuitry can often limit the ultimate *S/N*.

Adjacent channel power can result from high SSB phase noise of the transmitter's oscillator. It can be produced by the modulation method itself or result from rapid transmitter turn-on and turn-off. In narrowband systems, the amount of adjacent channel power integrated over the corresponding receiver's IF bandwidth is usually required to be more than 50 dB below carrier power. In wideband systems, 80 dB can be achieved.

Frequency stability is especially important in narrowband systems and is a closely regulated and monitored transmitter parameter.

Intermodulation distortion is produced when strong external signals are picked by the transmitter's antenna and the mixing products are then retransmitted, causing interference in other systems. One or more circulators can be placed between the last active stage and the antenna to reduce transmitter IM distortion. Environmental conditions and reliability issues are just as important as in receiver design. The transmitter's dc power supply must supply the required power over voltage and environmental variations.

1.2.1 Transmitter Architecture

A typical transmitter block diagram is shown in Figure 1.14. Frequency of operation, modulation method, and other details of the actual application dictate most of the architecture decisions, but some generalizations are possible.

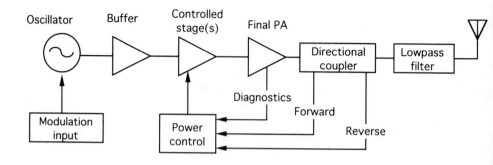

Figure 1.14 Typical transmitter architecture.

Amplitude (AM) and angle (FM and PM) modulation methods require differ-ent transmitter architectures. In an FM transmitter, the oscillator is modulated; in AM applications, modulating the oscillator requires linear amplifier stages through-out the chain. Another more efficient AM method involves modulating the DC supply of the final device (which can be a class-C amplifier stage) to produce amplitude modulation. The modulator must be an audio power amplifier, in order to supply the amperes of current required by the high-power RF stage.

Transmitters can have fixed or variable power output level. A transmitter whose power output can be varied over a wide range is inherently more difficult to design than a fixed-power circuit. Continuous-duty versus intermittent-duty cycle, operating bandwidth, temperature, and power supply variations are some of the other transmitter design considerations.

The purpose of the low-pass filter at the output in Figure 1.14 is to attenuate harmonics of the transmitted signal. This sounds simple but turns out to be more involved in actual application. The theoretical harmonic attenuation of the low-pass filter is almost never realized. Most low-pass filters are reflective; they achieve attenuation by impedance mismatch and reflection. This means that the undesired harmonics are reflected back into the device and remixed. This effectively increases the level of harmonic output from the active device, so the low-pass filter appears to lose selectivity. Therefore, the low-pass filter should be overdesigned to provide about 20 dB more rejection than would appear to be required.

The amount of amplification required from the buffer stages is primarily a function of their available input power and required output power. In transmitters where load pull needs to be minimized, more buffer amplifier stages with attenuators between them need to be used than required by straightforward gain requirements in order to increase the isolation between the oscillator and its changing load. Using additional buffer amplifiers and attenuators will cause the wideband noise to increase.

The relative position of the directional coupler, or power sensor, and the low-pass filter may be important in transmitters required to operate over a wide range of output powers. The coupling factor in the directional coupler must be sufficiently high to reliably sample the amplifier output at the lowest power setting, without being so tight as to allow harmonics generated by the detector diodes to couple back into the output at the highest power setting. The directional coupler may need to be followed by a low-pass filter to attenuate these harmonics. The disadvantage of such an arrangement is that the power-leveling loop cannot compensate for the low-pass filter frequency response and loss. The power sampling directional coupler should not be placed right at the output of a high-power stage. Its output is usually rich in harmonics, which may be preferentially coupled to the detector, depending on the directional coupler implementation, resulting in inaccurate power indication at the fundamental frequency.

The diagnostic signals in a high-power power amplifier (PA) usually include the final device current, temperature, and controlled stage voltage drive level, as well as the forward and reverse power levels. Many feedback loops are present in a high-power PA to control and sense all these parameters; care must be taken not to introduce low-frequency instability into the control loops. The power-control loop is frequently used to deliberately slow down the transmitter turn-on and turn-off times to minimize generation of spectral components on adjacent channels. This is particularly important in data communications, where the transmitter may be turned on and off in rapid succession as data bursts are being transmitted and received. Sections 1.2 and 1.4 and (1.22) show that this process generates a broad spectrum of energy, whose bandwidth can be minimized by deliberately slowing down the power ramp up and ramp down. As a rule of thumb, rise and fall times longer than 6 ms will produce energy components more than 70 dB down 25 kHz away from the main carrier.

Similar to the receiver case, the antenna must operate at dc ground to avoid damage from static charges. In simplex transceivers, there is usually a transmit/receive (T/R) switching network to connect the antenna to the transmitter or receiver, as desired. Section 2.18 illustrates some common T/R switching topologies.

The main challenge in the modulation circuitry is to provide constant modulation (either FM deviation or percentage AM, as the case may be) over a broad RF frequency range, as well as over wide baseband frequency range. Digital signals, which may contain significant low frequency as well as dc components, present a particular challenge in FM modulation of synthesized signal sources. Since the phase-locked loop will correct any slowly changing frequency shift imposed on the VCO, the reference frequency must be modulated as well. The converse is true for the reference source: its frequency can be changed at a slow rate, but fast frequency changes will not propagate through the loop filter and thus will never modulate the VCO output. Therefore, the low-frequency components are fed to the reference, and high modulating frequencies are fed to the VCO, in exact balance with respect to amplitude and phase, to produce constant frequency deviation for all modulating frequencies.

1.2.2 Load Pull

Load pull refers to the tendency of the transmitter frequency to change as the power output builds up. There are four reasons for this frequency shift.

As the power output ramps up, the device impedances change; when the effect of these impedance changes propagates backward to the oscillator, its load impedance changes, causing its frequency to shift. The cure for this problem is to provide enough buffering to isolate the oscillator from any load impedance variations. Keep in mind that a buffer stage that operates in amplitude compression has reduced reverse isolation. A chain of class-A buffers with attenuators in between is one method of improving isolation over a broad frequency range. Another technique for dealing with load pull suitable for synthesized transmitters involves widening the phase-locked loop (PLL) bandwidth for a short period of time as the transmitter is keyed. The wider loop bandwidth allows faster frequency correction to be applied, compensating for any frequency shift imposed by a changing load.

Sudden requirement for a large dc current may momentarily put a dip on the system power supply voltage, affecting every system that does not have its own dc supply regulator. VCOs in particular need a well-filtered and regulated power supply.

If the oscillator is not shielded from the transmitter, or if RF energy is allowed to flow back to the oscillator through the power supply leads, grounds, or control lines, the phase relationships in a phase-locked loop will be momentarily upset, because the phase of the high-power signal may not be the same as that already present. An extreme case of rectification of the high-level signal may result in dc bias and operating-point changes in the oscillator.

Changing magnetic field from increasing bias current flowing into the PA will induce voltages in coils and long circuit board traces. The rise time of this field is relatively long, and conventional shielding may not be effective in protecting sensitive circuitry. The thickness of a metal shield should be several skin depths to shield effectively. For example, the skin depth in copper for a rise time of 0.25 ms is over 2 mm. Ferromagnetic (steel) shields are more effective than copper against changing magnetic fields.

1.3 CABLE DISTRIBUTION AND COMMUNICATION

The most familiar cable communication system is cable television, but other communication systems in tunnels, mines, and buildings use leaky cable as the transmission medium. High-speed computer networks and some telephone trunk systems also use coaxial cable. Many of these systems are gradually converting to fiber optics whenever extremely wide operating bandwidths are required. Virtually all cable communication systems use unbalanced coaxial cable to take advantage of its shielding properties. Some cable system specifications follow.

Cable attenuation is the first specification, especially as it changes with operating frequency. Cable systems frequently use 75Ω as the characteristic impedance, because cables in this impedance range have the lowest attenuation. Cable attenuation at 300 MHz ranges from 8 dB/100 ft for thin cable to less than 1 dB/100 ft in specially made large-diameter cable. Cable attenuation increases with temperature approximately 0.2% per degree Celsius.

Cable voltage standing wave ratio (VSWR) causes mismatch loss and thus increases the effective attenuation. Signal reflections will cause ghosting in a CATV system.

Cable leakage may be desirable (e.g., in tunnel communication) or undesirable (e.g., in CATV).

Velocity of propagation is a concern mainly in computer communication networks; cables with very little dielectric material inside are used to keep the propagation velocity as high as possible. Cables made for this purpose usually have a narrow teflon strip helically wound around the center conductor so that most of the dielectric is air. The characteristic impedance of such cables tends to be high, near 100Ω.

S/N ratio is driven by two conflicting requirements: the signal amplitude should be kept as high as possible to reduce the number of cable repeaters, while low signal amplitudes are advantageous because less IM distortion is generated.

Cross-modulation involves the transfer of modulation from one carrier to another in a multicarrier system.

1.3.1 Cable Repeaters

The two main concerns in cable distribution are sufficient *S/N* ratio at all distribution points and predictable IM products resulting from cable repeaters. A cable distribution network can be conveniently analyzed if the repeater amplifiers' gains are assumed to be exactly equal to the interconnecting cable losses; see Figure 1.15. The overall noise figure of this chain can be calculated from (1.20).

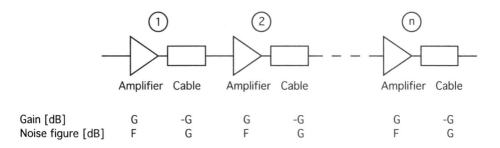

	Amplifier	Cable	Amplifier	Cable	Amplifier	Cable
Gain [dB]	G	-G	G	-G	G	-G
Noise figure [dB]	F	G	F	G	F	G

Figure 1.15 Cable repeater chain of identical sections.

$$F_T = 10 \, \log\left[1 + n\left(F - \frac{1}{G}\right)\right] \qquad (1.20)$$

When the amplifier gain and number of repeater links are large, the following approximation, found in the literature [7], can be used:

$$F_T \approx 10 \, \log(F) + 10 \, \log(n)$$

F_T = system equivalent input noise figure (dB)

F = noise factor of each amplifier (all assumed equal) (linear)

n = number of repeater amplifiers in system

G = amplifier gain, reciprocal of cable loss (linear)

The system noise figure increases by approximately 3 dB each time the number of amplifiers doubles. There is usually an upper limit on the number of amplifiers in the system to maintain the output S/N ratio above a certain threshold, depending on the bandwidth of the cable channels.

The equivalent input IP3 can also be calculated for the amplifier/cable cascade. The total equivalent input intercept is equal to the output intercept, because the overall gain of the system is 0 dB. In this analysis, we assume that there is no IM in the passive cable sections. The equivalent intercept point also degrades by 3 dB every time the number of amplifiers doubles.

$$\text{IP3}_T = \text{IP3} - 10 \, \log(n) \qquad (1.21)$$

IP3_T = total equivalent input or output intercept point (dBm)

IP3 = input intercept point of each amplifier (all assumed equal) (dBm)

n = number of repeater amplifiers in system

Subsections 1.1.6.2 and 1.3.2 on intercept points and triple beats show that intermodulation products generated by third-order nonlinearities are basically of two types: two-signal IM and three-signal IM. Based on the mathematics for the two IM products, we can make the following observations regarding a system where carriers of equal amplitude are equally spaced in frequency and every channel is occupied. Interference produced by two-signal IM products is constant across the band, while interference produced by three-signal IM (triple beats) products is not constant, but peaks in the middle of the band. There is far more interference from three-signal IM products than from two-signal IM products, and each three-signal IM product is 6 dB higher than a two-signal product.

Figure 1.16 is a graphic representation of the number of two-signal and three-signal IM interference products resulting from ten equally spaced carriers. There

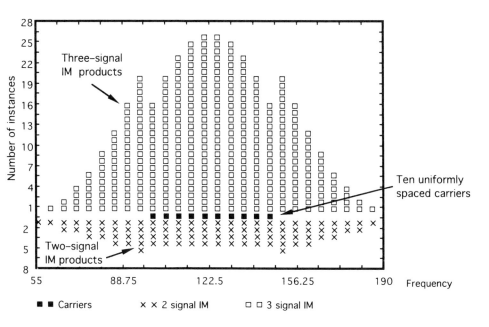

Figure 1.16 Distribution of IM products in a 10-carrier system [8].

are 26 instances of triple beats and only four instances of two-signal IM in the middle of the carrier band. Therefore, carriers in the middle of the band will suffer from lower S/N ratio than carriers near the edge, assuming that IM products rather than noise are the limiting S/N factor.

An interesting related challenge is the requirement of filling a particular band with carriers without causing any IM products to fall on any of the carrier frequencies. The basic rule of thumb to follow is that the spacing between any two carrier frequencies must not be the same. The plot in Figure 1.17 illustrates such a frequency allocation for eight carriers [8].

The eight carrier frequencies in Figure 1.17 are 100, 101, 104, 110, 121, 129, 134, and 136 MHz. No third-order IM products fall on any of the channels using such carrier placement. All such schemes will be more densely populated by carriers at the band edges, with the band center relatively free of carriers. Needless to say, such frequency allocations are not very efficient in terms of spectrum utilization, but many cable distribution systems can still benefit from thoughtful carrier placement.

Table 1.5 shows how to place up to nine carrier frequencies in order to avoid third-order IM products on any of the channels. The right column specifies the frequency increments in terms of channel spacings, to be added to the lowest frequency in generating the other carrier frequencies.

If we allow a maximum of one triple-beat occurrence per channel, the eight-carrier allocation can be changed to 0, 1, 9, 15, 21, 25, 26, and 28, resulting in better spectrum utilization.

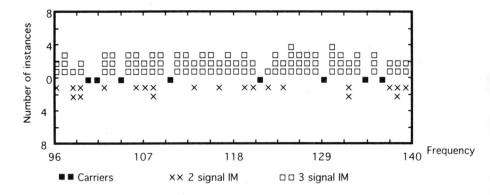

Figure 1.17 Placement scheme for eight carriers without IM interference.

Table 1.5
Frequency Allocation Without Third-Order IM Interference

Number of Carriers	Δf From Lowest, in Terms of Channel Spacings
1	0
2	0, 1
3	0, 1, 3
4	0, 1, 3, 7
5	0, 1, 4, 9, 11
6	0, 1, 4, 10, 15, 17
7	0, 1, 4, 10, 18, 23, 25
8	0, 1, 4, 10, 21, 29, 34, 36
9	0, 1, 4, 10, 21, 35, 43, 48, 50

Nonlinearities are not constrained to only third order. Even with clever frequency allocation, fifth- and higher-order nonlinearities will start limiting the S/N ratio in many-carrier systems.

1.3.2 Triple Beats

Triple-beat distortion is used to evaluate third-order mixing products resulting from relationship of the type $f_1 + f_2 - f_3$. Interference caused by triple beats is 6 dB higher than two-tone IM if all carriers have the same amplitude. Figure 1.18 shows all the possible third-order interactions among three carriers, f_1, f_2 and f_3, that fall near the band of interest. Additional triple beats will be produced near the third harmonic.

The number of triple beats resulting from a relationship of the type $f_1 + f_2 - f_3$ among n carriers is

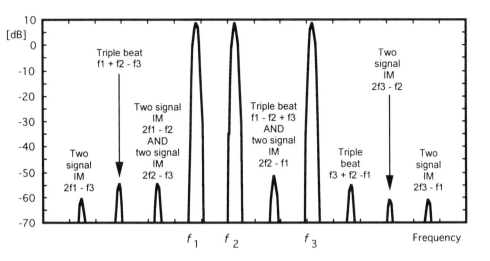

Figure 1.18 Third-order products resulting from three carriers of equal amplitude.

$$\text{Triple beats} = \frac{n(n-1)(n-2)}{2}$$

In multicarrier systems, triple beats are far more numerous than two-signal IM. Not only is the probability of three-signal IM distortion higher, but the IM product itself is of 6-dB higher amplitude than two-signal IM. Cable TV distribution systems and busy on-air environments are two examples where three-signal IM (triple beats) must be carefully taken into account in system design.

The composite triple-beat measurement starts with an analysis of the system to determine which channel has the highest potential number of triple-beat products within its bandwidth. In a system where the carriers are uniformly spaced and where every channel is occupied, the largest buildup of triple beats occurs in the middle of the signal band, as shown in Figure 1.16. In a system not populated by carriers at uniform frequency spacings, such as a commercial cable TV system, a proper system analysis must be undertaken to determine the frequency of maximum triple-beat buildup. Reference [9] contains a table of triple beats expected in a conventional cable TV system populated by seven different frequency allocations.

Once we know which channel occupies the worst location, we note the power level at that frequency when the channel is on and the power level when it is off. The composite triple-beat measurement in decibels below carrier is the difference between the two levels, as shown in Figure 1.19. A bandpass filter just for that channel may be required in the measurement in order not to overload the measurement instrument, usually a spectrum analyzer.

As triple beats accumulate in the middle channels, cable TV companies sometimes boost the signal levels there to preserve S/N ratio.

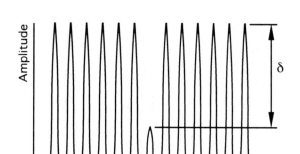

Figure 1.19 Composite triple-beat (=δ) measurement.

1.3.3 Bidirectional Cable Communication

Bidirectional cable communication can be used for routing separate signals in both directions on the same cable or to provide communication capability among portable, hand-held equipment in tunnels, mines, and buildings. These two modes are fundamentally different: in the first case, the forward and reverse signals can use separate frequency bands, while in the second case, the forward and reverse signals occupy the same frequency band. Two separate cables carrying communication traffic in opposite directions is the most straightforward, but not necessarily the cheapest, means of achieving bidirectional communication.

Bidirectional CATV communication (such as pay TV and monitoring systems) is an example of the first case, where separate frequency bands can be used to achieve bidirectional cable communication, as shown in Figure 1.20. All cable repeaters must be bidirectional amplifiers. The forward and reverse gains are usually different, because cable losses are higher at the higher frequencies.

The challenge of carrying signals that occupy the same frequencies in both directions can be approached in at least three ways. *Commutation* can be used to

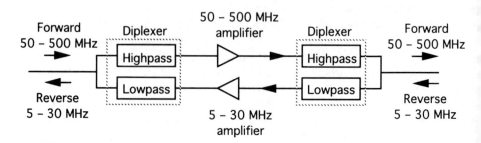

Figure 1.20 Bidirectional system using two separate frequency bands.

switch all cable repeaters from the forward to the reverse direction. Voice communication in the band between 300 Hz and 3 kHz requires that the switching rate be either below 300 Hz or above 3 kHz in order not to create in-band audible buzz. Cable propagation delays must be carefully analyzed, and for long cable runs the repeater switching should be staggered (not all switching at the same time).

Another way is to separate the forward and reverse communications in frequency by employing converter units at both ends of the cable, shown schematically in Figure 1.21. This is a system using leaky cable to provide communication capability among UHF portable radios. Communication in the forward direction (to the right) is achieved normally through UHF cable repeaters. Communication in the reverse direction is still to the right until the end of the cable where UHF traffic is downconverted to VHF and fed back to the base (to the left), where it is upconverted to UHF and sent down the cable again. The upconversion process must provide sufficiently high loss to avoid oscillation and echoes.

1.4 DATA COMMUNICATION

The perennial tradeoff in digital communication is BER versus transmission bandwidth. This is actually the same tradeoff as between analog S/N and channel bandwidth, just using different terminology. Low BER communications require broader transmission bandwidths for a given power level. If transmission bandwidth is not a concern, then frequency shift keying (FSK) and differential phase-shift keying (DPSK) are the modulation methods of choice. Both modulation and demodulation methods are easy, and low BER for a given S/N ratio is obtained (lower BER for DPSK). If transmission bandwidth is to be as narrow as possible, then multiple-level modulation methods involving both amplitude and phase must be used.

Quadrature phase-shift keying (QPSK) represents a good compromise between BER and transmission bandwidth, because it achieves virtually the same BER as coherent phase-shift keying (PSK) but uses only half the bandwidth. Many different variations of QPSK [10] (such as differential quadrature phase-shift keying [DQPSK], which has about 2.3 dB S/N penalty over coherent QPSK for the same BER) have been developed and used in digital communication products, such as modems.

There are two additional concerns with spectrally efficient multilevel modulation methods: (1) the occupied bandwidth required to transmit the signal with a minimum of errors and (2) interference caused on the adjacent channels by low-level spectral components from the main transmission channel. Digital and RF engineers may have some disagreements on this topic: modulation energy 50 dB down on the adjacent channel seems excellent to digital engineers, yet analog RF engineers know that so much power on the adjacent channel will cause significant system performance problems. Analog voice systems typically suppress adjacent channel feedthrough by more than 70 dB. If you are designing a digital communica-

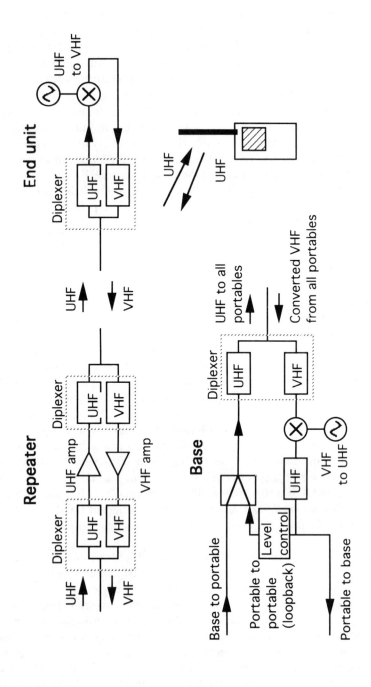

Figure 1.21 Bidirectional system using converters at both ends of the cable.

tion system, pay attention to adjacent channel requirements. Your digital transmitter may prevent authorized use of adjacent channels out to a wide radius around the transmitting site.

BERs for selected digital modulation methods [11] are determined from the following formulas.

On/off keying:

$$P_e = 0.5 \text{ erfc}\left[0.5\sqrt{(S/N)}\right]$$

Noncoherent amplitude shift keying (ASK):

$$P_e = 0.5e^{-S/(4N)} + 0.25 \text{ erfc}\left(\sqrt{\frac{S}{4N}}\right)$$

The probability of error is different for a mark than for a space. The above formula assumes that the message contains equal numbers of marks and spaces, that the decision threshold amplitude between a mark and a space is halfway between the two, and that we are operating at a high S/N ratio. Consult [11] for the optimum threshold level for low S/N ratios.

Coherent PSK, approximate for QPSK:

$$P_e = 0.5 \text{ erfc}\left[\sqrt{(S/N)}\right]$$

DPSK:

$$P_e = 0.5e^{-S/N}$$

DQPSK:

$$P_e \approx 0.5 \text{ erfc}\left[0.76\sqrt{(S/N)}\right]$$

FSK:

$$P_e = 0.5e^{-S/(2N)}$$

Coherent ASK:

$$P_e = 0.5 \text{ erfc}\left[\sqrt{S/(4N)}\right]$$

Coherent FSK:

$$P_e = 0.5 \text{ erfc}\left[\sqrt{S/(2N)}\right]$$

where

P_e = bit error probability

S = signal power (W) or energy per bit (J)

N = noise power (W) or noise density (W/Hz)

erfc = complementary error function (see Section 4.3)

The spectrum of a pulse train is

$$\text{Amplitude spectrum} = \sum_{n=1}^{\infty} 2\,A\,Tf\left[\frac{\sin(n\,\pi\,Tf)}{n\,\pi\,Tf}\right] \qquad (1.22)$$

n = harmonic of f, $n \neq 0$

A = pulse amplitude, bipolar (V)

T = pulse duration (s)

f = pulse repetition rate (Hz)

When an RF carrier is keyed on and off by such pulse train, the spectrum in (1.22) is translated in frequency up to the RF carrier frequency. This type of modulation is not spectrally efficient because the $\sin(x)/x$ function falls off very slowly (6 dB/octave) with frequency. When measuring the BER of RF links or their susceptibility to interference, it is sometimes advantageous to use a pseudo-random digital sequence as a test signal. The simple circuit shown in Figure 1.22 can generate such a pseudo-noise sequence of length 32,767 bits before repeating:

Other combinations of shift register length and feedback locations are possible [11,14] to generate even longer sequences.

Spread-spectrum techniques are basically of two types: direct sequence (DS) and frequency hopping (FH). Transmitted signals all occupy the same bandwidth, which is typically much wider than the message itself requires.

In DS spread spectrum, all messages are modified by different pseudo-random spreading codes such that all undesired messages look like noise. The desired message can be extracted using the original spreading code. Several rules of thumb should be observed in spread-spectrum communications.

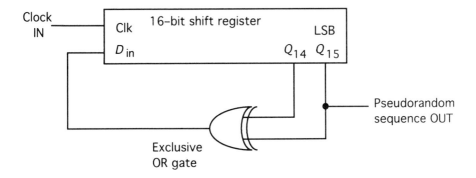

Figure 1.22 Pseudo–random sequence generator.

1. The spreading code rate must be much higher than the binary message rate.
2. A particular pseudo-random spreading code sequence must have high auto-correlation properties with itself and low cross-correlation properties with other possible codes, so that other encoded messages look like noise.
3. The spreading code must be statistically independent of the message. The encoded message spectrum must be noiselike as well.

Spreading codes can be generated by circuits similar to Figure 1.22 and are sometimes called Gold codes [12]. The reception of DS spread-spectrum signals relies on correlating the received signal with the desired message's known code [13].

The trend in spread-spectrum communications is toward dedicated ICs that perform the required spreading-code generation, message encoding, clock recovery, synchronization, and signal processing. Nevertheless, spread-spectrum systems have properties of which RF engineers need to be aware. For example, spread-spectrum systems are notoriously inefficient in spectrum utilization, not only because the transmitted bandwidth is much wider than the required message bandwidth, but because the spread spectrum itself has a $\sin(x)/x$ shape and has significant spectral components outside the band allocated to spread-spectrum communications. If you are operating in a frequency band adjacent to that allocated for spread-spectrum usage, expect your channel to contain much nonthermal "noise."

The spectrum of DS spread-spectrum signals is not truly noiselike but contains discrete components whose amplitude varies randomly. The frequency spacing of these discrete spectral components is equal to the rate at which the spreading code repeats itself (low kHz or less). The first null in the $\sin(x)/x$ spectrum will be at an offset equal to the code clock rate (MHz range). Thus, the overall spectrum shape is not determined by the message but is entirely given by the spreading code properties.

Reference [14] contains an example for the design of a spread-spectrum system with a specific required amount of continuous-wave (CW) interference rejection.

REFERENCES

[1] Meehan, M. D., and J. Purviance, *Yield and Reliability in Microwave Circuit and System Design.* Norwood, MA: Artech House, 1993.

[2] Smith, P. G., and D. G. Reinertsen, *Developing Products in Half the Time.* New York: Van Nostrand Reinhold, 1991.

[3] Carlson, A. B., *Communication Systems,* 2d ed. New York: McGraw-Hill, 1975, p. 446.

[4] Imagine That, Inc., 6830 Via Del Oro, Suite 230, San Jose, CA 95119, USA, (408) 365-0305.

[5] Vizmuller, P., *Filters With Helical and Folded Helical Resonators,* Dedham, MA: Artech House, 1987, p. 61.

[6] Vizmuller, P. "Two-Generator Method Improves Spurious Response Measurments," *Mobile Radio Technology,* August 1985, p. 60.

[7] Lynes, K., *CATV Systems Technical Manual, Design and Measurement.* Private publication, 1984.

[8] Courtesy of ingSOFT Limited, 213 Dunview Ave., Willowdale, ONT M2N 4H9, Canada, 416-730-9611, using the software package *RFIntercept.*

[9] Bartlett, E. R., *Cable Television Technology and Operations.* New York: McGraw-Hill, 1990, p. 178.

[10] Stremler, F. G., *Introduction to Communication Systems.* Reading, MA: Addison-Wesley Publishing, 1982, Chap. 10.

[11] Erst, S. J., *Electronics Equations Handbook.* Blue Ridge Summit, PA: Tab Books, 1989, pp. 99, 150.

[12] Moser, R., and J. Stover, "Generation of Pseudo-Random Sequences for Spread Spectrum Systems," *Microwave Journal,* May 1985, p. 287.

[13] Ziemer, R. E., and R. L. Peterson, *Digital Communications and Spread Spectrum Systems.* New York: Macmillan Publishing, 1985.

[14] Young, P. H., *Electronic Communication Techniques,* 2d ed. Columbus, OH: Merrill Publishing, 1990, pp. 760–764.

CHAPTER 2

Circuit Examples

2.1 ACTIVE FILTERS

The main advantage of active filters at low frequencies is that inductors are not required in their implementation. At higher frequencies, inductors are much smaller, and their use does not pose as much of a problem. Very high Q passive resonators are readily available at higher frequencies, while Q-multiplication by an active circuit is not reliable at high frequencies, because it may result in circuit oscillation due to wide variation in device properties over frequency and temperature and from unit to unit.

The main advantages of passive filters are their immunity from intermodulation (IM) distortion and their simple tradeoff between insertion loss and physical size; the intercept points and noise figures of active devices are not as easy to change to suit system requirements. Nevertheless, some active circuits are quite suitable for implementing basic filter topologies.

Distributed amplifiers are inherently designed as low-pass filters. The cutoff frequency is given by (2.1). FET input capacitance is used as a filter component (see Subsection 2.2.3).

$$f_c = \frac{1}{\pi R_0 C_{IN}} \tag{2.1}$$

f_c = upper cutoff frequency (Hz)

R_0 = system characteristic impedance, also characteristic impedance

of the equivalent input and output transmission lines (Ω)

C_{IN} = FET device gate to source capacitance (F)

The same basic structure can be used to implement Butterworth, Chebyshev, or linear phase filters. The disadvantage is that twice as many components are

---

53

required as when passive filters are used, because the basic filter structure must be repeated in both the input and output transmission lines of a distributed amplifier. Individually tuned amplifiers can also be cascaded with the proper coupling coefficients to form bandpass filters [1]. This configuration generally requires greater design effort than a simple cascade of a passive filter and amplifier, because of stability and linearity considerations.

Operational amplifier designs are constantly climbing up in frequency, and it is possible to design reliable high-frequency active filters using low-frequency techniques up to several MHz [2], but at a fairly high cost.

2.2 AMPLIFIERS

2.2.1 Single-Stage Amplifiers

The design of small-signal single-stage amplifiers begins with an examination of the device s-parameters and noise parameters to determine if that particular device is approximately suitable for the application. The stability k-factor should be calculated not only at the frequencies of interest but also out of band to determine potential instability as a function of frequency. Many amplifier designers skip this step, only to discover much later that major design effort is required to ensure stability, especially if temperature, power supply, load, and source variations are to be taken into account. Careful study of stability is mandatory for low-noise amplifiers directly connected to an antenna. Many antennas (especially those for hand-held devices) present an uncontrolled impedance environment to the amplifier input.

The key concepts in stabilizing a potentially unstable device ($k < 1$) can be summarized as follows: a purely reactive termination at either the input or the output of the device will not alter potential instability; only resistive attenuation can effectively improve circuit stability. Series or shunt resistors at either input or output can be used to achieve unconditional stability, but you should avoid using resistors at device input, in order not to degrade the noise figure (out-of-band resistive termination at device input is acceptable). The selection of either series or shunt resistor at device output will depend on the amount of gain reduction or additional mismatch produced by the stability fix; computer simulation can greatly help in examining different tradeoffs. Resistive feedback can also sometimes be used to stabilize an amplifier stage.

Series or shunt resistive feedback is often used to flatten the gain over frequency in broadband amplifier designs. Any stability analysis must include the effect of the feedback. It should be noted that feedback may affect the reverse isolation, an important amplifier property in many applications.

Once a stable topology is obtained, the device is either matched for low noise operation or for simultaneous conjugate match at input and output.

The small-signal schematic in Figure 2.1 illustrates all these concepts for a low-current amplifier, using the MRF931 device operating at 1 mA. Low-current

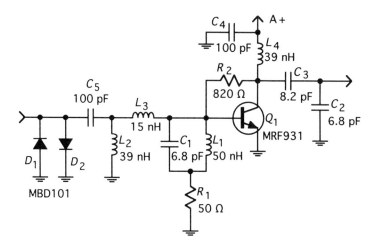

Figure 2.1 Single-stage amplifier with input diplexer to ensure stability (bias circuit not shown).

amplifiers are sometimes unstable because the input and output circuits are usually resonant circuits so that maximum gain can be obtained. Figure 2.1 shows a 10-dB amplifier operating at 300 MHz.

C_1 and L_1 form a parallel resonance near 300 MHz, so that the 50-Ω resistor is out of the circuit at 300 MHz but is present at all other frequencies, ensuring stability. This stability fix is placed close to the base of the device, so that the input matching network will not affect it at higher frequencies. The collector impedance is a resonant circuit tapped down to 50 Ω using a capacitive divider.

The two diodes D_1 and D_2 at the input provide burnout protection if the amplifier operates in an environment with potentially high signal levels present. This situation is rather common in practice; signal levels in excess of +20 dBm can result from many situations. Two cars can be parked side by side, and a keyed 25-W mobile transmitter in one car will overload the nearby receiver. A portable radio-telephone transmitter operating in close proximity to another receiver can cause problems. When several base stations share a common antenna, the signal amplitudes at the different frequencies and locations must be carefully determined. A keyed police motorcycle-mounted transmitter passing your car on the highway will overload your receiver.

If an RF amplifier is subjected to high input RF levels for an extended period of time, its noise figure will start degrading long before the device is actually destroyed. Therefore, a noise figure test is the most meaningful measurement in testing RF amplifiers for susceptibility to high input levels. The drawback of using two back-to-back diodes is that they can generate IM distortion when high input levels are present, even when the RF amplifier is shut off and powered down. Two diodes in series can be used to raise the threshold clipping level. Emitter impedance is generally undesirable because it causes instability, but sometimes a small emitter impedance is deliberately used to improve the third-order intercept of a single-stage RF amplifier. Figure 2.2 illustrates the IP3 improvement, gain reduction, and k-factor degradation for an MRF580 at 50-mA bias with emitter impedance.

The gain decreases, intercept point improves, and stability degrades with increasing emitter impedance. All these effects are more pronounced at lower frequencies, because internal emitter inductance diminishes the influence of external components at higher frequencies.

In low-noise, high-intercept-point designs, the emitter impedance is provided by a small external inductor, because a resistor will degrade the noise figure. The worst scenario for stability happens when this external inductance forms a parallel resonance with the inevitable stray capacitance, providing an infinite emitter impedance at some very high frequency. In that case, oscillation is almost guaranteed. The topology shown in Figure 2.3 can be used to provide attenuation above and below the operating frequency to limit the available gain and ensure stability.

L_1 and C_1 are series-resonant at the operating frequency, effectively shunting R_1. The L_1/C_1 ratio will determine the bandwidth of this network. Since it provides attenuation at lower frequencies as well, it can be used to flatten the gain by resonating L_1 and C_1 slightly above the operating frequency band.

2.2.2 Multistage Amplifiers

Amplifier stability is of key importance in the design of RF and microwave amplifiers. All the stability criteria that we have at our disposal apply only to single-stage

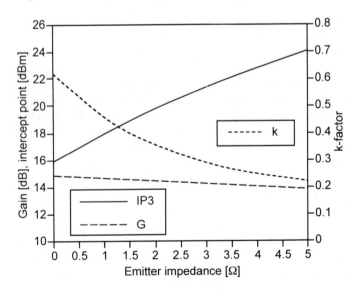

Figure 2.2 Variation of VHF amplifier properties with emitter impedance.

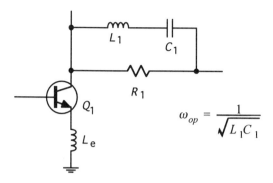

Figure 2.3 Collector circuit for improving out-of-band stability.

amplifiers. Therefore, when designing an amplifier cascade, make sure that every single amplifier stage is stable on its own, even if stability analysis of the whole cascade indicates that its k-factor is greater than 1.

In multistage amplifiers, the low-noise stage should be first, and the highest intercept stage should be last. If overall gain control is required, the gain of the first stage is usually controlled to avoid overload under strong signal conditions. In many applications, the control of stage gain by changing its bias current is unacceptable, because to obtain lower gain a lower bias current is required; this action lowers the intercept point, which is undesirable, because lower gain is required exactly when input signal levels are high. Clearly, with high input signal levels we would instead like to operate at a higher bias current to avoid distortion. In such applications, electronic attenuators are a much better choice for controlling gain (see Section 2.3).

The total amount of gain that is achievable with multistage amplifiers is ultimately limited by the isolation between output and input. If the overall gain is greater than the isolation, oscillation may result. Coupling through power supply and control lines has to be taken into account when considering isolation between output and input.

Multistage amplifiers are most frequently used in transmitter driver circuits; in most such applications, low isolation between output and input can cause load-pull, which is a change in frequency as the transmitter power ramps up. Section 1.2 addressed this concern in more detail.

2.2.3 Distributed Amplifiers

Distributed amplifiers operate differently from conventional amplifiers, where the device is usually matched such that its input resistance absorbs as much power from the source as possible. In distributed amplifiers, the device input and output capacitances are used in a lumped-element approximation to a transmission line, so that the amplification is accomplished by a true transconductance function: device output current is proportional to voltage appearing across its input capacitance.

The input capacitance of a FET is usually higher than its output capacitance, and additional discrete capacitors, C_a, must be used in parallel with each transistor output to maintain the same phase velocity in the input and output transmission lines, as shown in Figure 2.4. Otherwise, the traveling waves will not add up for maximum gain. Note that both input and output transmission lines are properly terminated in their characteristic impedances, R_0.

The two critical design parameters are the transistors' C_{IN} and g_m, the gate to source capacitance and device transconductance.

$$C_a = C_{IN} - C_{OUT}$$

$$A_V = 0.5(-n\ g_m R_0) \tag{2.2}$$

$$f_c = \frac{1}{\pi R_0 C_{IN}}$$

$$L = C_{IN} R_0^2$$

C_a = additional output line capacitance (see Figure 2.4) (F)

C_{IN} = FET device input, gate to source capacitance, includes
   Miller effect capacitance (F)

C_{OUT} = FET device output, drain to source capacitance (F)

A_V = low-frequency voltage gain (linear)

n = number of transistor stages

g_m = FET device transconductance (mho, A/V)

R_0 = system characteristic impedance, also characteristic impedance of
   the equivalent input and output transmission lines (Ω)

f_c = upper cutoff frequency (Hz)

L = series inductance in the lumped equivalent input and
   output transmission lines (H)

The distributed amplifier is a classical case of gain-bandwidth tradeoff. For example, choosing a high R_0 we can achieve higher gain but for a proportionally narrower bandwidth. Some general guidelines apply to this class of amplifiers:

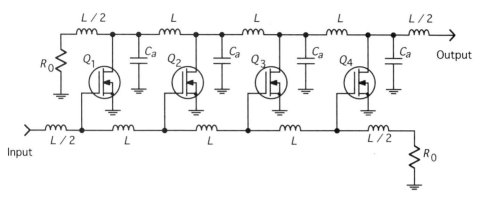

Figure 2.4 Four–stage distributed amplifier.

1. The device transconductance should be quite high for useful amplification.
2. Input capacitance should be low for the widest operating bandwidth.
3. There are no inherent low-frequency limitations; with proper biasing, amplification down to DC is possible. Keep in mind, however, that this is an inverting amplifier, and a dual-polarity power supply may be required to accommodate positive and negative voltage swings.
4. Feedback capacitance from drain to gate must be very low, otherwise its Miller equivalent will increase the apparent C_{IN} and will reduce the operating bandwidth.

$$C_{IN} = C_{gs} + C_{gd}(1 + 0.5g_m R_0)$$

5. Because of device imperfections, additional series inductance in the drain lead will sometimes improve real-life performance.
6. Bipolar transistors are not suitable for this type of amplifier, because their input impedance is not predominantly capacitive, unless their bias current is very low.
7. Intercept point can be improved by increasing the bias currents of the devices close to the output of the device chain.

The distributed amplifier concept can also be expanded into two dimensions to form a two-dimensional matrix of transconductance amplifiers.

2.2.4 Low-Current Cascode Amplifiers

Low-current drain applications impose unique design constraints on RF amplifier designs. Bipolar transistor impedances increase, the gain drops, and tendency toward instability increases. Low-power applications are usually concerned with current drain, while the available voltage is usually above 3V, as required by other (probably digital) circuitry. The cascode connection has the advantage that both stages share the same bias current, so that higher gain and better reverse isolation can be achieved without any increased power drain penalty. A 3-V supply is still adequate for biasing both transistors.

Figure 2.5 shows a 300-MHz amplifier with 18 dB of gain and 38 dB of reverse isolation, operating from 3V at 0.5 mA. The 680-Ω collector resistor required for stability also lowers the output Q for wider operating bandwidth. The IM performance of low-current amplifiers is rather poor.

2.2.5 Power Amplifiers

Linear amplifiers are designed mostly for conjugate match at both input and output ports. Class-C power amplifiers are usually conjugately matched only at their input. The output circuit is usually mismatched, because the load resistance is determined by the desired output power:

$$R_L \approx \frac{(V_{CC} - V_{sat})^2}{2\,P_0} \tag{2.3}$$

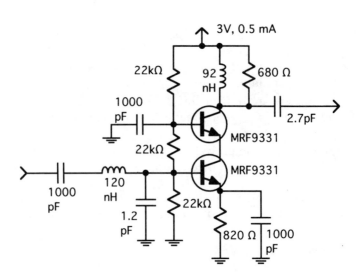

Figure 2.5 Low-current cascode amplifier.

R_L = load resistance required at device output (Ω)

V_{CC} = DC supply voltage (V)

V_{sat} = device saturation voltage (V)

P_O = desired RF power output (W)

Equation (2.3) is valid for Class A, B, or C assuming that the device's own output impedance can be neglected. This assumption becomes invalid at high frequencies, in which case step 1 of the following design procedure has to be modified as described later in this section.

The tasks in designing a power amplifier in a frequency region where (2.3) applies are:

1. Design a matching network that transforms the system impedance (usually 50 or 75 Ω) to the required load resistance, R_L, over the required bandwidth.

2. Design a conjugate match at the device input when the output is terminated in the required R_L.

3. Verify that the amplifier is stable over temperature and voltage variations and output voltage standing wave ratio (VSWR). This last task usually takes the most time. The most sensitive component affecting stability is the collector (or drain) dc choke. The impedance of this dc feed circuit must not be too large at low frequencies.

4. Verify that the device voltage, current and power dissipation ratings are not exceeded over the operating conditions. The collector or drain voltage in class-C amplifiers will reach twice the supply voltage.

Figure 2.6 illustrates techniques for improving power amplifier stability. L_1 and L_2 are RF blocks at operating frequency. L_3 is typically much larger than L_2, and R_1 is low, around 10 Ω. $C_1 < C_2$, and C_1 is an RF bypass at operating frequency. R_1 ensures low collector impedance at low frequencies, where the active device has plenty of gain. E_1 is a ferrite bead included to introduce loss into the base circuit at low frequencies.

A more general design technique, which does not rely on (2.3), requires accurate measurement of low impedances: A double-stub tuner is placed between the device and the desired load; the tuner is adjusted for whatever performance criteria are desired: maximum power output, high efficiency, low IM distortion, and so on. The impedance presented to the device output lead is measured over the required frequency and supply voltage range. This required impedance is then synthesized by means of discrete components and transmission line sections. This same technique can also be used to identify impedances that result in undesirable operation, such as instability, excessive current drain, even self-destruction. Such impedance regions are then deliberately avoided in the synthesis procedure.

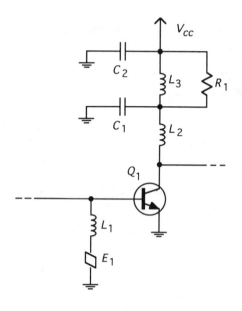

Figure 2.6 Base and collector circuits to ensure stability.

The first difficulty associated with this technique is the requirement for accurate measurement of low impedances. Conventional network analyzers lose accuracy near the edge of the Smith Chart, and therefore direct impedance measurement of the required load impedance requires great care in minimizing stray inductances. A useful technique for minimizing this difficulty is to include a fixed impedance step-up from the device to the tuner [3]. This technique effectively increases the resolution of the double-stub tuner at low impedances and reduces uncertainties associated with fixture and tuner losses at high VSWRs.

The second and more serious difficulty comes in estimating device-to-device variations in the required load impedance. Many power amplifier designers can vividly recall instances of their designs suddenly not meeting specifications, because a new batch of devices with different date codes appeared in the factory. Often the device manufacturers themselves cannot predict the anticipated device variations, and the designer is left with no choice but to design in a very conservative manner, to leave some reserve and not push devices to their limits.

Termination impedance presented to the device at harmonics of the operating frequency is also of interest. The operating efficiency can often be improved by providing an open circuit termination at odd harmonics, and short circuit termination at even harmonics of the operating frequency. Such terminations can be realized by use of a shorted transmission line dc feed, which is a quarter-wave long at the operating frequency.

High-power amplifiers frequently require cooling by means of air flow. The fan capacity required to maintain a certain thermal rise when dissipating a certain amount of power can be estimated.

$$\Phi \approx 1.8 \frac{P}{\Delta T}$$

Φ = air flow (ft³/minute)

P = power generated (W)

ΔT = allowed temperature rise (°C)

Cooling without airflow has to take into account all the thermal resistances between device junction and ambient environment [4]. The aim is not to exceed a certain junction temperature T_j required for reliable operation.

$$T_j = P_d \, \theta_{ja} + T_a$$

where

$$\theta_{ja} = \theta_{jc} + \theta_{cs} + \theta_{sa}$$

$$P_d = P_{DC} + P_{in} \pm P_{out} \text{ (for explanation of sign, see below)}$$

T_j = junction temperature (°C)

P_d = power dissipated in junction (W)

P_{DC} = input dc power (W)

P_{in} = input RF power (W)

P_{out} = output RF power (W). Sign is minus for matched operation, plus for operation into infinite VSWR, such as would occur during a fault, or if antenna were disconnected or badly mismatched.

θ_{ja} = junction to ambient thermal resistance (°C/W)

T_a = ambient temperature (°C)

θ_{jc} = junction to case thermal resistance (°C/W)

θ_{cs} = case to heat sink thermal resistance (°C/W)

θ_{sa} = heat sink to ambient thermal resistance (°C/W)

Power amplifier designs using FET power transistors instead of bipolar power transistors are gaining more and more popularity because of the greater stability of the power FET when operating into an impedance mismatch. The input impedance of FET is also more constant with drive level than bipolar input impedance.

2.3 ATTENUATORS, MINIMUM LOSS PADS

Attenuators are used to adjust signal levels, to control impedance mismatch, and to isolate circuit stages. Important attenuator properties are attenuation flatness with frequency, VSWR, and power-handling capability. The familiar pi-circuit and T-circuit attenuator topologies are shown in Figure 2.7 and Figure 2.8.

Pi-circuit component values:

$$R_3 = \frac{1}{2}(10^{L/10} - 1)\sqrt{\frac{Z_{in}Z_{out}}{10^{L/10}}} \tag{2.4}$$

$$R_2 = \frac{1}{\dfrac{10^{L/10} + 1}{Z_{out}(10^{L/10} - 1)} - \dfrac{1}{R_3}} \tag{2.5}$$

$$R_1 = \frac{1}{\dfrac{10^{L/10} + 1}{Z_{in}(10^{L/10} - 1)} - \dfrac{1}{R_3}} \tag{2.6}$$

Z_{in} = desired input resistance (Ω)

Z_{out} = desired output resistance (Ω)

L = transducer loss (dB)

Minimum loss that can be achieved for given Z_{in} and Z_{out}:

$$L_{min} = 20 \log\left(\sqrt{\frac{Z_{in}}{Z_{out}}} + \sqrt{\frac{Z_{in}}{Z_{out}} - 1}\right) \tag{2.7}$$

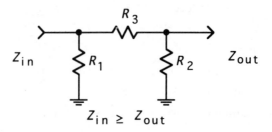

Figure 2.7 Pi–circuit attenuator topology.

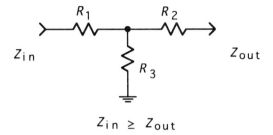

$$Z_{in} \geq Z_{out}$$

Figure 2.8 T–circuit attenuator topology.

T-circuit component values:

$$R_3 = \frac{2\sqrt{Z_{in}Z_{out}10^{L/10}}}{10^{L/10} - 1} \tag{2.8}$$

$$R_2 = \frac{10^{L/10} + 1}{10^{L/10} - 1}Z_{out} - R_3 \tag{2.9}$$

$$R_1 = \frac{10^{L/10} + 1}{10^{L/10} - 1}Z_{in} - R_3 \tag{2.10}$$

Z_{in} = desired input resistance (Ω)

Z_{out} = desired output resistance (Ω)

L = transducer loss (dB)

Minimum loss that can be achieved for given Z_{in} and Z_{out}:

$$L_{min} = 20 \log\left(\sqrt{\frac{Z_{in}}{Z_{out}}} + \sqrt{\frac{Z_{in}}{Z_{out}} - 1}\right) \tag{2.11}$$

Bridged-T attenuator topology is shown in Figure 2.9.

$$R_1 = Z_0(10^{L/20} - 1) \tag{2.12}$$

$$R_4 = \frac{Z_0}{(10^{L/20} - 1)} \tag{2.13}$$

L = desired attenuator loss (dB)

Z_0 = terminating, system impedance (Ω)

The bridged-T topology is frequently used with PIN diode electronically con-trolled attenuators, as shown in Figure 2.10, because only two diodes need to be

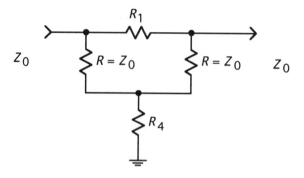

Figure 2.9 Bridged–T attenuator topology.

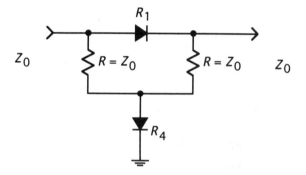

Figure 2.10 Variable attenuator using PIN diodes.

controlled (as opposed to three with the T-circuit or pi-circuit configuration). Variable attenuators using PIN diodes can be designed to have less intercept-point variation with attenuation setting than a functionally comparable transistor automatic gain control (AGC) stage.

Proper attenuator operation in Figure 2.11 requires that the currents through D_1 and D_4 vary inversely to each other: At low attenuations, current through D_1 must be high, and current through D_4 must be low. Conversely, at high attenuator settings, current through D_1 must be low and through D_4 high. This is accomplished by having D_1 and D_4 share the same dc load, R_6. Since the anode of D_1 is fed from a fixed voltage, increasing current through D_4 forces the voltage across R_6 to rise, thus decreasing the current through R_5 and D_1. Therefore, as the analog control voltage increases, the current through D_4 increases, current through D_1 decreases, and attenuation goes up. The exact resistor and voltage values must be determined from the particular PIN diode characteristics.

If resistor R_5, R_6, and R_7 values in the pi attenuator in Figure 2.12 are high (greater than 1 kΩ), then the corresponding chokes, L_2, L_1, and L_3, need not be

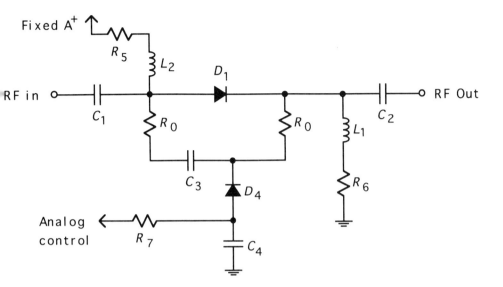

Figure 2.11 A dc bias arrangement for bridged–T electronic attenuator.

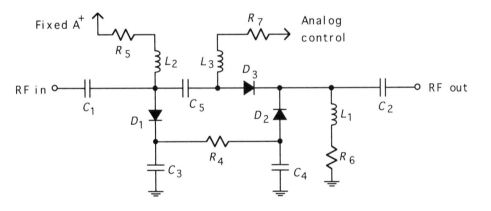

Figure 2.12 A dc bias for a pi-circuit variable electronic attenuator.

used. The principle of operation of the pi attenuator is the same as the bridged-T attenuator: analog control voltage sets the current through R_7 and D_3. Since D_3 and the series connection of D_1, D_2 share a common dc load, R_6, the currents through D_3 and series D_1, D_2 vary inversely to each other, as required.

The pi attenuator in Figure 2.12 has an interesting advantage over the bridged-T attenuator in Figure 2.11 in AGC applications, because the first diode in the signal path D_1 has a high bias current at high attenuation settings. This has the advantage of generating less IM distortion, because high attenuator settings are typically required when signal levels are high (high diode current results in better IM performance). The situation of having low bias currents at high signal levels, and hence high attenuator settings, is to be avoided. This undesirable situation is all too common in AGC amplifiers, where the gain is lowered by decreasing the bias current at the expense of IM distortion exactly at the conditions that require good IM performance.

PIN diode attenuators have a lower frequency limit of operation, related to the transit time. The insertion loss increases and IM distortion becomes progressively worse at low frequencies. A double-balanced mixer can be used as a current-controlled attenuator at lower frequencies. The RF port is the input, the LO port is output, and the IF port is the dc current control [5]. The minimum attenuation obtainable is about 2 dB, and such an attenuator may not be well matched throughout its attenuation range; its intercept point is usually higher than that of a corresponding PIN diode attenuator at lower frequencies.

If an attenuator operates between a source and a load of unequal impedances, the definition of attenuation is actually taken to be transducer loss (see the example in Section 4.5) and is equal to the ratio between power absorbed by the load and the power available from the source.

High-power, low-value attenuators can be constructed from lengths of transmission line. One application of such attenuators is in testing high-power amplifiers for stability into a mismatch, as shown in Figure 2.13.

If the lossy transmission line has attenuation "A" dB, then the return loss presented to the device under test is "2A" dB. The adjustable-length line simply changes the phase of the reflection, to check stability for a full circle on the Smith Chart. To check for stability into a 3:1 mismatch, 3-dB attenuation is required from the lossy line. To construct the balanced pi or T attenuators in Figure 2.14 and Figure 2.15, the series resistors need to be one-half of their unbalanced values; the shunt resistors remain unchanged.

Equations (2.4) through (2.7) also apply for designing balanced pi-circuit attenuators. Use Figure 2.14 to determine which resistor values are halved.

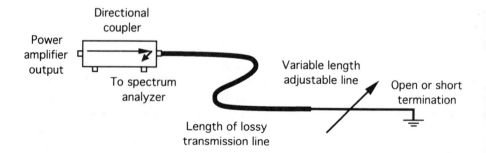

Figure 2.13 Length of lossy line used as low-value attenuator to test amplifier stability.

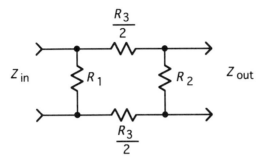

Figure 2.14 Balanced pi–circuit attenuator.

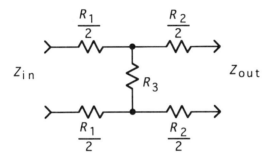

Figure 2.15 Balanced T–circuit attenuator.

Equations (2.8) through (2.11) still apply for designing balanced T-circuit attenuators (also called H-circuit attenuators). Use Figure 2.15 to determine which resistor values are halved. For example, Figure 2.16 shows a 10-dB, 300-Ω balanced attenuator derived from (2.8) through (2.10) and Figure 2.15.

2.4 BALUNS AND TRANSFORMERS

Many circuits can be designed with different topologies to operate in one of two modes, balanced or unbalanced, distinguished from each other by the function of

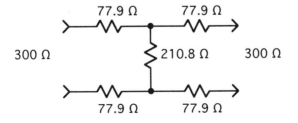

Figure 2.16 Balanced 10-dB, 300-Ω attenuator.

ground. Balanced circuits require ground at a voltage potential halfway between the terminals; the ground usually carries no signal current. Unbalanced circuits use ground as the return path for signal current. TV twin lead is an example of a balanced transmission line; coaxial cable is an unbalanced transmission line. Many devices perform better in one or the other mode, requiring a *balun* (*bal*anced to *un*balanced) device to connect the two modes of operation with low loss and minimal impedance mismatch. Typical balanced circuits are mixers, lattice crystal filters, and antennas; the majority of other circuits are implemented in an unbalanced topology.

A simple *LC* circuit, which can perform the unbalanced to balanced function, is shown in Figure 2.17.

The input, represented by R_S, is unbalanced; the output, represented by R_L, is balanced.

$$L = \frac{\sqrt{R_S R_L}}{2 \pi f}$$

$$C = \frac{1}{2 \pi f \sqrt{R_S R_L}}$$

R_S = source resistance, unbalanced (Ω)

R_L = load resistance, balanced (Ω)

f = frequency of operation (Hz)

The basic relationships for a tapped coil, shown in Figure 2.18, are

$$L = L_1 + L_2 + 2M$$

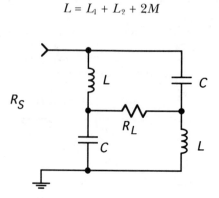

Figure 2.17 Narrowband bridge balun.

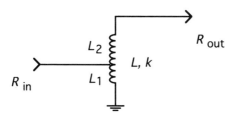

Figure 2.18 Tapped coil, or autotransformer.

$$tap = \frac{n_1}{n_1 + n_2} \times 100 = \frac{\sqrt{L_1}}{\sqrt{L_1} + \sqrt{L_2}} \times 100$$

$$k = \frac{M}{\sqrt{L_1 L_2}}$$

$$R_{in} \approx \left(\frac{tap}{100}\right)^2 R_{out} \qquad (2.14)$$

$$R_{out} \approx \left(\frac{100}{tap}\right)^2 R_{in}$$

subject to the conditions:

$$R_{in} \ll \omega L_1$$

$$R_{out} \ll \omega L$$

$$k \approx 1$$

L = total inductance between output and ground, with input open (H)

k = coupling coefficient between two windings, $0 < k < 1$

L_1 = inductance between input and ground, with output open (H)

L_2 = inductance between input and output, with "ground" unconnected (H)

M = mutual inductance between L_1 and L_2 (H)

n_1 = number of turns in L_1

n_2 = number of turns in L_2

tap = tap point, measured from ground (%)

ω = $2\pi f$, radian frequency (rad/s)

Since the conditions listed for a tapped coil are rarely achieved in practice, computer analysis and optimization should be applied when tapped coils are used as impedance transformers.

Transformers wound in the conventional manner of separate primary and secondary windings, each with their own number of turns, fail to perform their function at higher frequencies because of inadequate flux linkage due to the winding method used or loss of permeability of the ferromagnetic material and because of interwinding capacitance. The art of designing high-frequency transformers consists of including the interwinding capacitance in a transmission line structure, so that at high frequencies the structure can be analyzed as an interconnection of transmission lines, while it looks like an autotransformer at lower frequencies.

Many different transformer configurations are possible; the rest of this section only briefly describes the most common types, and the interested reader is encouraged to review [6–8] for additional information. Ferrite baluns with the best performance and widest operating bandwidth are wound with very fine coaxial cable rather than the usual multifilar wire strands.

The frequency response of transmission line transformers wound on ferrite cores will be governed by ferrite properties at low frequencies and $\lambda/4$ wire length at high frequencies.

Figure 2.19 shows the output balanced load isolated from the input unbalanced source; the actual voltages at the balanced terminals with respect to ground will be predominantly determined by parasitic capacitances. This situation may be undesirable in some cases, where we would like to not only provide isolation but also force balanced symmetrical operation with respect to circuit ground, as in driving push-pull amplifiers. The configurations in Figure 2.20 and Figure 2.21 provide this capability.

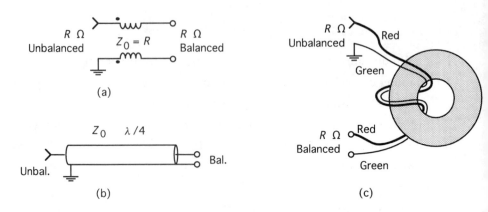

Figure 2.19 Simple 1:1 balun: (a) schematic diagram, (b) transmission line form, and (c) construction.

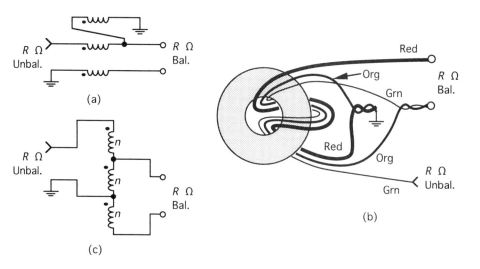

Figure 2.20 Series–aiding balun with isolation and symmetry (1:1): (a) schematic diagram, (b) construction, and (c) low-frequency equivalent circuit.

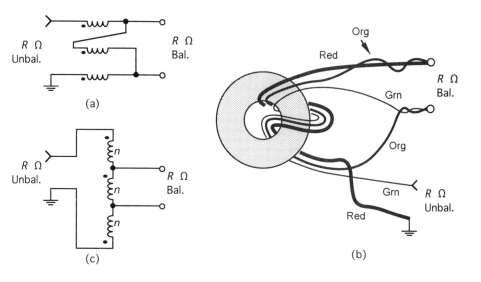

Figure 2.21 Series–canceling 1:1 balun: (a) schematic diagram, (b) construction, and (c) low-frequency equivalent circuit.

An impedance transformation is sometimes desirable in a balun. The transmission line circuit in Figure 2.22 can provide impedance transformation, depending on the bifilar line characteristic impedance. An additional balun of the type in Figure 2.19 can be added on either side for improved performance.

At low frequency, there is always a 4:1 impedance transformation; at high frequencies, the impedance transformation ratio depends on Z_0, the characteristic impedance of the bifilar winding. A 4:1 impedance transformation is achieved with $Z_0 = 2 \times R$, while a 1:1 high-frequency transformation results with $Z_0 = R$. Frequency-dependent impedance transformation can thus be achieved. This requirement may seem unusual, but it is often encountered in practice, because the characteristic impedance of bifilar wire in common use is near 50 Ω, rather than the 100 Ω required by a proper 4:1 transformation. Note that the configuration in Figure 2.22 must remain balanced at its output, meaning that neither output terminal should be connected to ground. A 4:1 impedance transformation between two unbalanced loads can be achieved using the autotransformer in Figure 2.23.

By exchanging the input and ground connections at the input only, a 4:1 unbalanced to balanced transformer results.

Baluns and impedance transformers that do not use ferrite transformers but rely only on transmission lines are suitable for UHF and microwave frequencies and are sometimes used even at lower frequencies, because their power-handling capability is much higher. Figure 2.24 shows the simplest such baluns with and without impedance transformation.

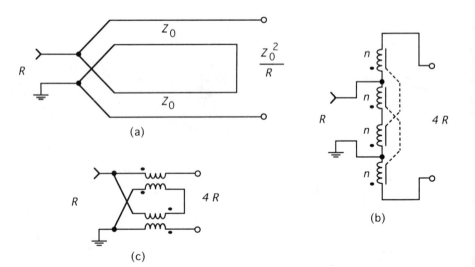

Figure 2.22 Balun with impedance transformation: (a) transmission line form, (b) low-frequency circuit, and (c) schematic diagram.

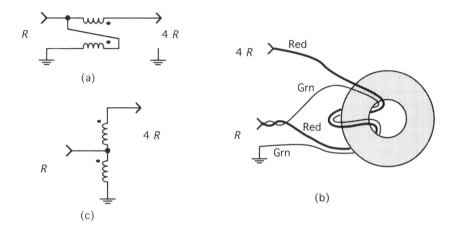

Figure 2.23 Unbalanced 4:1 transformer: (a) schematic diagram, (b) construction, and (c) low-frequency circuit.

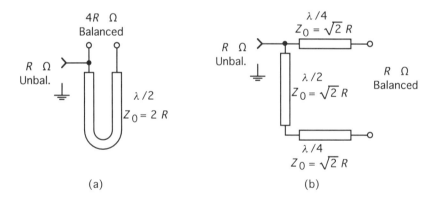

Figure 2.24 The $\lambda/2$ transmission line baluns: (a) 4:1 transformation, and (b) 1:1 balun.

The basic configuration in Figure 2.19(a) can be approximately realized by two coupled lines in microstrip, shown in Figure 2.25, with the disadvantage that an impedance step-up from balanced to unbalanced side must be accommodated, because the odd mode impedance of edge-coupled microstrip lines is usually quite high. Since the even and odd mode velocities are different in microstrip coupled lines, the frequencies of best VSWR and best isolation will not coincide.

More complicated baluns can be constructed by using the collinear or Marchand topology, as illustrated by the coaxial implementation in Figure 2.26.

Decade bandwidth is possible with the Marchand balun [9], with or without impedance transformation. Reference [10] contains a wealth of additional informa-

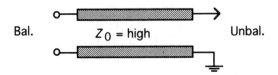

Figure 2.25 Microstrip balun.

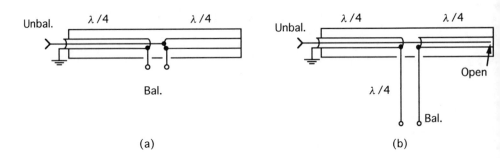

Figure 2.26 Cross section of two coaxial baluns: (a) collinear, and (b) Marchand.

tion on transmission line and ferrite transformers suitable for high-power circuits below 200 MHz.

2.5 BIAS NETWORKS

Class-A common-emitter amplifiers are usually very sensitive to stray impedance in the emitter circuit. Any small inductance in series with the emitter will cause instability; for this reason the emitter needs to be grounded as directly as possible, and bias components in the emitter are generally undesirable. In the schematic in Figure 2.27, Q_1 is the RF amplifier, and Q_2 provides its base current required for constant voltage difference across R_c. This constant voltage difference then ensures constant collector current.

Diode D_1 provides some measure of temperature compensation. R_b should be high in order not to affect base impedance, but not high enough to cause Q_2 to saturate over temperature and β_1 variations. Neglecting the base current of Q_2, the design equations are

$$I_c = \frac{R_1(A^+ - V_d)}{R_c(R_1 + R_2)}$$

$$V_c = A^+ - I_c R_c$$

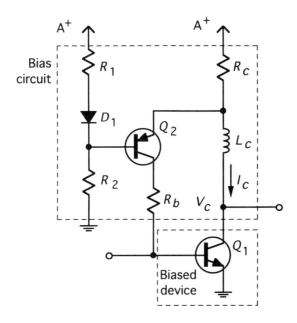

Figure 2.27 Active bias network for a common–emitter RF amplifier stage.

Assuming that we are designing the bias circuit to provide a certain device bias current I_c and collector voltage V_c, select a convenient $A^+ > V_c$. The component values are then supplied by the following equations:

$$R_c = \frac{A^+ - V_c}{I_c}$$

$$R_1 = \frac{A^+ - V_c}{I_d}$$

$$R_2 = \frac{V_c - V_d}{I_d}$$

$$R_b < \beta_{min} \frac{V_c - V_d - 0.2}{I_c}$$

I_c = desired collector current of Q_1 (A)

V_c = desired collector voltage of Q_1 (V)

V_d = diode, or base-emitter voltage drop, nominally 0.7 (V)

A^+ = chosen supply voltage (V)

R_i = resistor values as shown in Figure 2.27 (Ω)

I_d = bias current through R_1, R_2, and D_1 (A)

β_{min} = minimum beta of Q_1

The bias circuit shown has to be carefully bypassed at both high and low frequencies. There is one inversion from base to collector of Q_1, and another inversion may be introduced by L_c, matching components and stray capacitances, resulting in positive feedback around the loop at low frequencies. Low *ESR* electrolytic or tantalum capacitor from the collector of Q_2 to ground is usually adequate to ensure stability.

2.6 CRYSTAL AND CERAMIC FILTERS

Crystal filters are used whenever very narrow bandpass filters are required, such as in receiver IF stages for adjacent channel rejection, in SSB systems for sideband rejection, and following oscillator stages for noise attenuation. They are low-frequency devices, limited to less than 200 MHz. The primary design difficulties lie in their high input and output impedances, low tolerance to overload, susceptibility to IM distortion, and the presence of undesired responses. This section deals primarily with crystal filters, but filters utilizing ceramic resonators can be treated in a similar manner.

2.6.1 Monolithic Filters

Monolithic crystal filters are usually three-terminal, two-pole devices with very closely controlled center frequency and bandwidth. They are most frequently used in the IF stages of a receiver. The input and output impedances are resistive and range from hundreds to thousands of ohms. Therefore, some kind of impedance-transforming network is always required; the large impedance transformation ratio requires careful attention to matching network component value tolerances and value changes over temperature.

Stray capacitances associated with the particular circuit board layout will be important because the impedances involved are quite high. Monolithic filter specifications usually do not include the statistical variation of the terminal impedances, so designers may need to include tunable components in the matching and coupling networks. This situation is improving, and nontunable matching networks are becoming more prevalent.

Intermodulation performance can be a problem; many receivers require IF filters selected for good IM performance. Often there will be pronounced variation in IM distortion as a function of frequency offset from center frequency. Furthermore, monolithic filters suffer from a peculiar memory effect; when subjected to overload, the intercept point degrades and takes a long time (minutes or longer) to recover. Very high signal levels in excess of 0 dBm can permanently damage the resonators.

Cascaded monolithic filters require careful attention to terminal impedance variation from unit to unit, which will affect the overall bandwidth. The filters have internal capacitances to ground from the input and output pins, which need to be absorbed by the external circuitry. A 6-dB attenuator is sometimes used between cascaded monolithic filters to minimize their impedance interactions.

Spurious responses on the high-frequency side of quartz crystal filters can reduce spurious response rejection, especially the second image.

Figure 2.28 shows a typical monolithic filter matching network topology assuming an impedance step-down from the filter to the external environment. The possibility of operating at a higher reference impedance than 50 Ω should always be considered, because the matching problem will become easier.

Components C_{10}, C_{11}, and L_1 are the input impedance match. When cascading monolithic filters, a capacitor from the junction of the two filters to ground is usually adequate; for critical applications, a 6-dB pad consisting of R_1, R_2, and R_3 together with separate C_4 and C_5 do a better job at avoiding filter interaction due to tolerance accumulations. The resistive pad is also sometimes used to reduce ringing in the time domain when the filter is subjected to pulsed RF signals and to reduce passband ripple in the frequency response. A small valued capacitor C_3, sometimes called a *gimmick capacitor*, introduces two transmission zeros on either side of the passband for steeper selectivity, if required. The penalty for this is reduced selectivity past the transmission zero frequencies.

2.6.2 Ladder Filters

Most of the filtering requirements in a typical receiver can be satisfied by monolithic filters; nevertheless, there are situations that require special filter designs. Single-sideband systems, comb filter banks, and electronic instruments have particular narrowband filter requirements. Narrow crystal front-end filters at dense base station antenna sites are frequently the only alternative available for interference-free receiver operation. Ladder crystal filters can also be designed for wider bandwidths than monolithic crystal filters. Theoretical filter design using resonators coupled to each other by impedance inverters is well understood; in such designs, all the

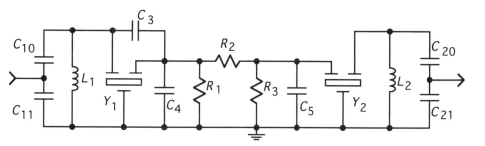

Figure 2.28 Cascaded monolithic crystal filter topology.

resonators are tuned to the same frequency. Since ideal impedance inverters do not yet exist, their approximate realization forces the series resonators (crystals) to be of different frequencies. This is a major drawback of ladder crystal filters, because crystals of different frequencies are required to implement a filter of more than two poles. Figure 2.29 shows a theoretical design that uses crystals of the same frequency coupled to each other by impedance inverter approximations. The key consideration is that the coupling inductors L_{12} and L_{23} cannot be physically realized with high enough Q, and they have to be absorbed by L_1, L_2, and L_3: $L_1' = L_1 + L_{12}$, $L_2' = L_2 + L_{12} + L_{23}$, and $L_3' = L_3 + L_{23}$. Absorbing the impedance inverter inductances into the resonator inductances makes the crystal resonant frequencies different. The crystal shunt capacitance may be neglected for very narrowband designs, but it must be included for wider bandwidths designs, because the transmission zero that it introduces above the passband will decrease and ultimately limit the bandwidth.

The design procedure can be summarized as follows:

1. Obtain the required coupling coefficients (k and q) for the desired number of poles and desired response shape (Butterworth, Chebyshev, etc.) using the methods in Section 2.10. It must be clearly understood here that the k and q coefficients for Chebyshev filters are different depending on whether the 3-dB or the ripple bandwidth is used for reference. Refer to Subsection 2.10.1 for further details.

2. Calculate impedance inverter characteristic impedance X.

3. Calculate coupling capacitor values C_{12} and C_{23}, new crystal resonant frequencies, and terminating impedances.

4. Design matching network for the desired input and output impedances.

5. Optimize the design to compensate for crystal shunt capacitance.

The procedure is best illustrated by an example, similar to the design in [11]. Suppose that we use four crystals whose motional capacitance is 0.012 pF, shunt package capacitance is 3.3 pF, and $Q = 60,000$ to design a filter centered at 10

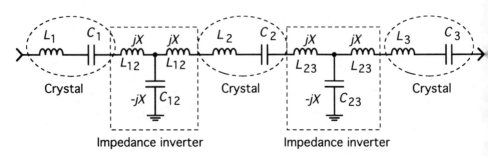

Figure 2.29 Ladder crystal filter prototype.

MHz, with a 3-dB bandwidth of 2 kHz and 0.1 dB ripple, intended to operate in a 50-Ω environment.

The coupling coefficients are determined by the filter type and the number of resonators; from (2.26) to (2.34): $q_1 = q_4 = 1.35$, $k_{12} = k_{34} = 0.69$, $k_{23} = 0.542$. Compute the following:

$$R_S = R_L = \frac{|X_1|BW_{3\,dB}}{q_1 f_0} = \frac{BW_{3\,dB}}{2\,\pi f_0 C_1 q_1 f_0} = \frac{2000}{2\,\pi\,0.012 \times 10^{-12}(1.35)\ 1 \times 10^{14}} = 196.5$$

$$X_{12} = X_{34} = \frac{\sqrt{X_1 X_2} k_{12} BW_{3\,dB}}{f_0} = \frac{k_{12} BW_{3\,dB}}{2\,\pi f_0 C_1 f_0} = \frac{0.69 \times 2000}{2\,\pi\,0.012 \times 10^{-12} \times 1 \times 10^{14}} = 183.0$$

$$X_{23} = \frac{\sqrt{X_2 X_3}\ k_{23} BW_{3\,dB}}{f_0} = \frac{k_{23} BW_{3\,dB}}{2\,\pi f_0 C_1 f_0} = \frac{0.542 \times 2000}{2\,\pi\,0.012 \times 10^{-12} \times 1 \times 10^{14}} = 143.8$$

$$C_{12} = \frac{1}{2\,\pi f_0 X_{12}} = 87\ \text{pF},\ L_{12} = \frac{X_{12}}{2\,\pi f_0} = 2.9125\ \mu H$$

$$C_{23} = \frac{1}{2\,\pi f_0 X_{23}} = 110.7\ \text{pF},\ L_{23} = \frac{X_{23}}{2\,\pi f_0} = 2.2886\ \mu H$$

Considering that the nominal crystal motional inductance is 21.1086 mH, the new resonant frequencies can be computed by recalculating the resonator inductance when impedance inverter inductances L_{ij} are absorbed:

$$L'_1 = L'_4 = L_1 + L_{12} = 21.1115\ \text{mH}$$

$$L'_2 = L'_3 = L_2 + L_{12} + L_{23} = 21.1138\ \text{mH}$$

And the new crystal frequencies are 9.99931 MHz and 9.99876 MHz. The final task is to design an impedance transformation network using Figure 2.79 and (2.48). Let $C_1 = 1000$ pF, then $C_2 = 982$ pF and $L = 500$ nH. The final filter is shown in Figure 2.30.

In this particular example, the 3.3-pF shunt package capacitance narrows the bandwidth by placing a transmission zero on the high-frequency side of the passband. Further computer optimization may be used to fine-tune the design, if required. It should also be mentioned that very narrowband designs using all crystals of the same frequency are still feasible, with degradation of inband ripple and insertion loss. Band reject filters are also possible. Figure 2.31 shows the required topology with shunt capacitors required for narrowband designs.

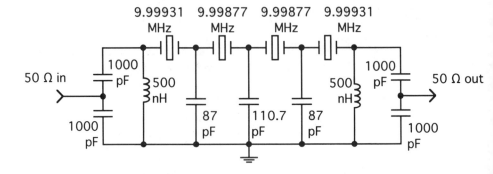

Figure 2.30 A 10-MHz crystal ladder filter.

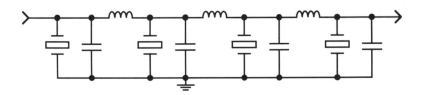

Figure 2.31 Narrowband crystal notch filter topology.

2.6.3 Lattice Filters

The bandwidth of a crystal ladder filter is affected at best, and limited at worst, by the crystal shunt capacitance. This is avoided in lattice type filters, which allow much wider bandwidths; the lattice topology is also more versatile in accommodating different filter requirements. The common approach in designing lattice filters is to start with a symmetrical ladder design, apply Bartlett's bisection theorem, and compensate the design for crystal shunt capacitance [11]. This procedure is outlined in Figure 2.32.

While such theoretical approach in principle yields the correct filter design, the filter performance will be affected by nonideal component performance, especially in the hybrid transformer. The best approach is to use a cascade of two-pole sections in which each section is optimized to compensate for nonideal components. Figure 2.33 shows the two-pole building block lattice filter. Note that crystals of different resonant frequencies are required to implement lattice filters as well.

Compared to the ladder type, lattice filters typically require higher terminating impedances and are more difficult to optimize, because all the component values interact. For example, using a different terminating impedance not only creates a mismatch, but also shifts the passband of the filter. Different shunt capacitance requires a different terminating impedance. The transformer properties are very

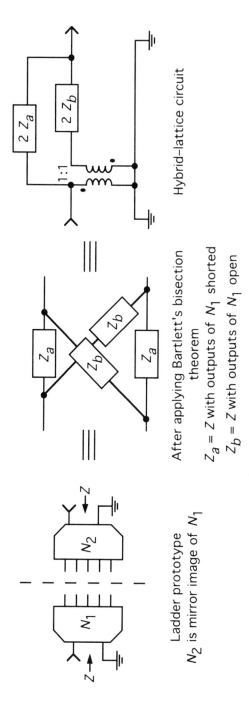

Figure 2.32 Derivation of basic hybrid–lattice circuit from a symmetrical ladder prototype.

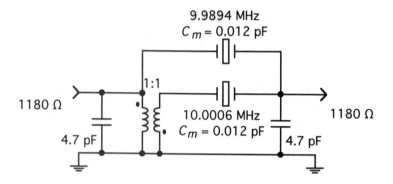

Figure 2.33 Two-pole hybrid–lattice filter of 12-kHz bandwidth at 10 MHz.

important as well; often the input capacitor in the transformer primary has to be higher in value to tune out transformer leakage inductance. All these interactions dictate that lattice filters be designed with tunable components; the filter is trimmed to the desired frequency response after assembly.

Figure 2.34 shows a cascade of two-pole sections to create a four-pole filter. Designs that use only one transformer are available, but they suffer from greater sensitivity to component value tolerances and place higher requirements on the transformer. Computer-aided circuit optimization is highly recommended to compensate for nonideal component properties.

The required input and output inductances can be supplied by the transformers.

2.7 DETECTORS AND MODULATORS

An RF carrier of constant amplitude and phase carries no information other than its presence. Amplitude modulation represents the simplest and historically earliest

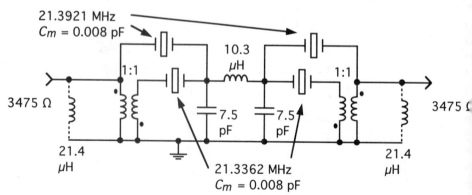

Figure 2.34 Four-pole hybrid-lattice filter of 80-kHz bandwidth at 21.4 MHz.

form of information encoding. Frequency and phase modulation have the benefit of constant amplitude, allowing the use of efficient class-C amplifiers in signal amplification. As the RF spectrum becomes increasingly crowded with signals, the technical community is faced with one of two basic choices: ever narrower bandwidth analog channels or completely digital forms of communication. In either case, precise control over modulation properties requires thorough familiarity with the different modulation and demodulation methods.

The most important property of any modulator or demodulator is its linearity: how faithfully it encodes or decodes the baseband signal without introducing undesirable distortion. A modulator's modulation sensitivity refers to the absolute level of baseband signal required to produce the desired amount of modulation. High modulation sensitivity makes a modulator susceptible to modulation by thermal noise, thus limiting the available signal to noise ratio, while low modulation sensitivity requires too much amplification of the baseband signal prior to modulation.

2.7.1 AM Detectors and Modulators

AM modulation can be accomplished by many means, including passive or active Gilbert cell mixers, electronic attenuators, or controlling the bias of an amplifier stage. Suppressed-carrier AM modulation is achieved by a double-balanced mixer, which is called *balanced modulator* when so used. Highly linear AM modulation can be accomplished by first using a balanced modulator to generate double-sideband suppressed carrier modulation and then adding a properly phased carrier of the correct amplitude to result in the desired AM modulation index.

The simplest AM detector is a half-wave rectifier used as a peak detector. Figure 2.35 shows four variations of this circuit: positive and negative peak detectors, voltage multiplier, and biased negative peak detector. Other variations include resistively terminated, matched, and prebiased circuits.

A continuous dc loop should exist through the diode and resistor R. This is the purpose of the choke L, which can be substituted by a resistor with some loss in sensitivity. If the dc return is absent and the peak detector is capacitively coupled to an RF source, the coupling capacitor will charge up to twice the maximum RF level and the diode will no longer conduct. Such a detector will still "detect" the presence of RF, but will not be able to follow amplitude variations. The detector's speed will depend on the amount of leakage present in the coupling capacitor and thus will not be well controlled. The RC time constant determines the maximum modulating frequency that can be received without distortion for a given modulation index. Consider an AM signal of high modulation index, where the envelope is changing very fast. If the RC time constant is too long, the demodulated signal will not be able to follow the trailing edges of a fast-changing envelope. The influence of the external circuit connected to the detector must be taken into account in this analysis.

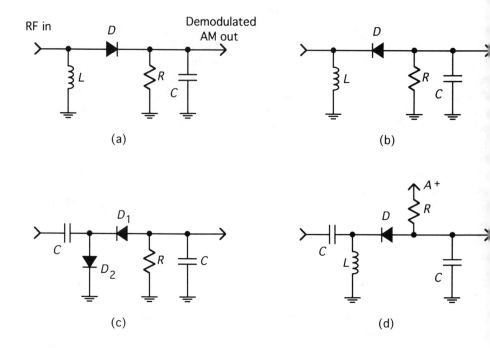

Figure 2.35 AM demodulator circuits using diodes. (a) Positive peak detector, (b) negative peak detector, (c) voltage multiplier, and (d) biased negative peak detector.

$$RC \le \frac{\sqrt{\dfrac{1}{m^2} - 1}}{3.8 f_{\text{max}}}$$

R = resistor value (Ω)

C = capacitor value (F)

m = modulation index, $0 \le m \le 1$

f_{max} = highest modulating frequency (Hz)

If the diode ON resistance is too high (as may happen at low input levels), then the rising edges of the waveform will be affected also, because the capacitor will not be able to charge fast enough through the high diode resistance. Special Schottky diodes, called zero-bias detectors have been developed to avoid this limitation. Forward biasing a regular Schottky diode with a small dc current will also help improve the detector response speed at low RF levels.

Since diodes are high-impedance devices, such a detector will be inherently mismatched to an RF source. A 50-Ω resistor is sometimes used in place of L to properly match the detector to a 50-Ω source. Biasing lowers the diode impedance, making impedance matching by reactive components somewhat easier.

Diode peak detectors are sensitive to temperature variations. This is especially true of biased detectors, where a change in output may signify the presence of an RF signal, or it may indicate a change in temperature!

Diode peak detectors typically require voltage swings near 1V to operate properly. Large system gain is required to amplify a weak incoming signal up to the levels required by the detector. However, when a strong carrier is received, some stages may be driven into compression, with the attendant loss of amplitude information. Therefore, all AM receivers need AGC to avoid clipping high-level signals. AGC is usually implemented by changing the dc bias of amplifier stages or by electronic attenuators. The average dc value of the peak detector output can be used to sense the amplitude of the incoming signal.

Schottky diode detectors operate as square-law devices (they detect power) at low levels, changing to peak voltage detectors at high input levels; their sensitivity and linearity thus change with input level.

Highly linear AM detection can be accomplished by synchronous demodulation using a mixer with its LO signal phase synchronized with the incoming AM signal. Such detectors are used for suppressed carrier and SSB demodulation. The synchronized carrier is usually recovered by a separate narrowband phase-locked loop, provided that some amount of carrier (pilot carrier) is always transmitted and is available for synchronization. Another method of carrier recovery relies on the generation of phase and frequency error signals in a multiple-feedback system [12] using two local oscillators in phase quadrature. Voice SSB systems are rather tolerant of phase drift, while high-speed data transmission requires careful phase synchronization of the recovered carrier.

The superregenerative detector (see Section 2.17) is often used for inexpensive detection of on-off keyed (OOK) signals.

2.7.2 FM Detectors and Modulators

FM modulation is most frequently accomplished by direct reactance modulation in the oscillator circuit. FM modulation of a stand-alone, inaccessible oscillator is called indirect FM and relies on the fact that narrowband frequency modulation is equivalent to phase modulation, if we first integrate the message prior to phase modulation.

The circuit in Figure 2.36 requires a dc bias on the modulation input line to bias varactor diode D_1. The same circuit topology can be used to FM-modulate surface acoustic wave (SAW) oscillators up to 1 GHz.

VCOs can be modulated in one of two ways: The steering line can be used, or a separate modulation circuit designed, especially for high-performance, wide-tuning VCOs. The steering line sensitivity is usually too high for well-controlled deviation control over a wide tuning range.

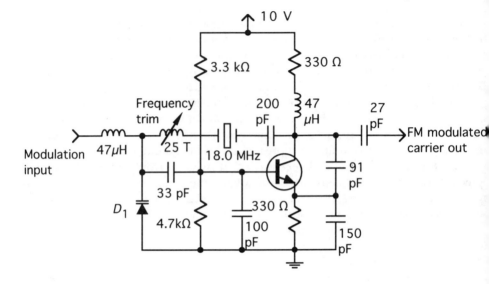

Figure 2.36 Direct FM modulation of an 18-MHz crystal oscillator.

Many types of FM demodulators have been developed. The following summary attempts to highlight the main features of each type. The theory of operation is covered in the references. The key performance specifications are detector sensitivity, *S/N* ratio required at detector input to produce the desired baseband *S/N* (sometimes called co-channel rejection), and bandwidth of operation, which will determine the maximum frequency deviation that can be demodulated.

The balanced slope detector [13] in Figure 2.37 is conceptually simple, requires two tuning adjustments, is suitable for wideband FM, and is easy to imple-

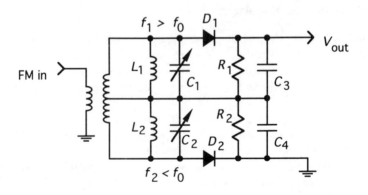

Figure 2.37 Balanced slope detector.

ment at microwave frequencies. One tuned circuit is slightly above carrier and the other one slightly below carrier frequency.

The crystal discriminator in Figure 2.38 is based on the Foster-Seely discriminator and uses a special piezoelectric crystal in place of the input transformer. It may need one tuning adjustment, only suitable for narrowband FM, and low IF frequencies, but it produces the highest demodulated output level for given frequency deviation input.

The delay-line FM demodulator in Figure 2.39 is the conceptual starting point for the Foster-Seely discriminator and is only useful for wideband FM. The main idea is that the first derivative of an FM signal contains the frequency as an amplitude term [14]: a differentiator converts FM to AM, which can then be detected by an envelope detector. A differentiator can be approximated by a short delay and 180° power combiner:

The Foster-Seely discriminator [13] in Figure 2.40 is used for narrowband FM systems, requires one or two tuning adjustments, and can achieve good sensitivity. Transformer primary and secondary are both tuned to the carrier frequency.

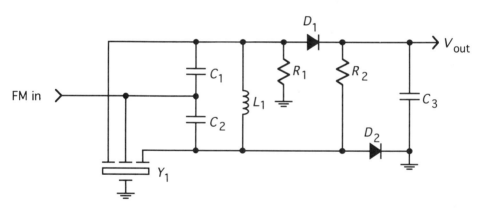

Figure 2.38 Crystal FM discriminator.

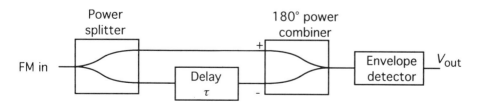

Figure 2.39 Delay line FM discriminator.

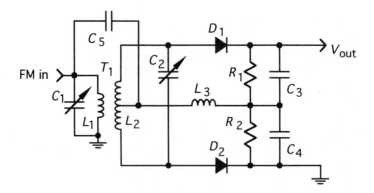

Figure 2.40 Foster–Seely discriminator.

The PLL demodulator in Figure 2.41 is suitable for wideband and narrowband FM demodulation, requires no tuning adjustments, and has inherent AM rejection. It is implemented in many integrated-circuit FM demodulators.

The quadrature detector [13] in Figure 2.42 is the favorite for FM receiver integrated circuits (no transformer required). It may require a tuning adjustment. The principle of operation relies on small-value capacitor C_1 and resonant circuit with controlled Q to provide quadrature signal to phase detector. Deviations from resonant frequency will produce deviations from phase quadrature, which can then be detected by a phase detector, usually implemented as an analog multiplier inside the IC.

The ratio detector [13] in Figure 2.43 is less sensitive than the Foster-Seely, but it has better AM rejection and requires one or two tuning adjustments. Voltage across C_6 absorbs AM variations and can be used for AGC, if required.

It has been mentioned that FM modulation of an oscillator whose internal circuit is not accessible may be difficult, yet exactly such a modulation method is offered by the Doppler effect in real life. We can simulate the Doppler shift by modulating the phase (i.e., time of travel) of an existing signal and rely on the relationship that frequency modulation is the time derivative of phase modulation.

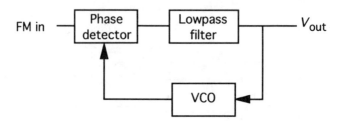

Figure 2.41 PLL FM demodulator.

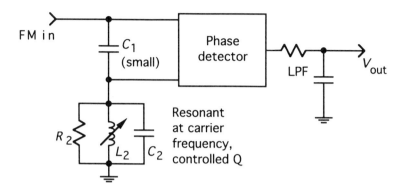

Figure 2.42 Quadrature detector.

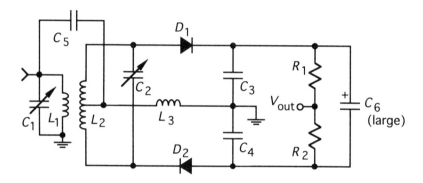

Figure 2.43 Ratio FM detector.

In such a system, the message would first be integrated and applied to a phase modulator. The resultant spectrum would then be equivalent to direct narrowband FM. Since the amount of phase modulation is usually limited, the attainable frequency deviation is far less than would be required in a practical system, and frequency multiplication may have to be used to obtain wider frequency deviations. There is a direct relationship between frequency multiplication and frequency deviation: If you double the frequency, you double the deviation.

The same relationship holds for a variant of FM: the SSB phase noise of an oscillator. If you double the frequency of an oscillator, you degrade its SSB phase noise by 6 dB. Frequency division similarly improves the SSB phase noise.

FM systems greatly benefit from pre-emphasis and de-emphasis, a technique that recognizes that FM demodulation introduces high-frequency noise into the demodulated signal. Prior to modulation, the message is processed to amplify high-frequency portions of the message (pre-emphasis). After demodulation, the message and the introduced noise are filtered to attenuate their high-frequency component (de-emphasis). Thus, the message is not altered, while high-frequency demodulate noise is suppressed. We must know how the signal has been pre-emphasized befor we can de-emphasize it in the receiver. Narrowband FM systems with 5-kHz frequenc deviation use 750-μs pre-emphasis, while wideband broadcast FM systems usuall take advantage of 50-μs, 75-μs (in North America), or 150-μs (in Europe) pre emphasis. Dolby compression systems use 25-μs pre-emphasis. The number in micr seconds, τ, refers to corner frequency, ω_1, in Figure 2.44, $\tau[s] = 1/\omega_1[rad/s]$.

$$\omega_1 = \frac{1}{\tau} = \frac{1}{R_1 C_1} = \frac{1}{R_3 C_3} \qquad (2.15$$

$$\omega_2 = \frac{1}{R_1 C_1} + \frac{1}{R_2 C_1}$$

ω_1 = pre-emphasis and de-emphasis corner frequency (rad/s)

τ = pre-emphasis and de-emphasis (s)

ω_2 = frequency well above message bandwidth (rad/s)

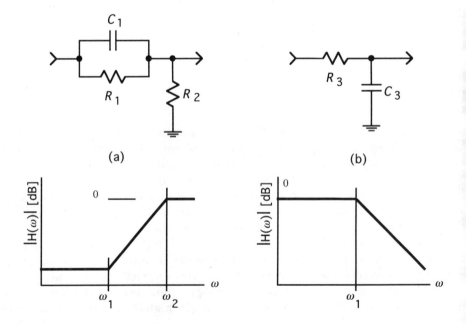

Figure 2.44 (a) Pre–emphasis and (b) de–emphasis.

The networks shown in Figure 2.44 assume that the source resistance is zero and the load resistance is infinite. If this is not the case, the source and load impedances have to be taken into account; if they are not known, buffer amplifiers or active filter implementations have to be used.

The pre-emphasis network frequency response increases with frequency and approximates a differentiator. We know that when a carrier is FM-modulated with the time derivative of the message, this is the same as phase modulation. Thus, most FM systems in use today are a combination of frequency modulation (FM) for low modulating frequencies and phase modulation (PM) for high modulating frequencies to take advantage of the best characteristics of both modulation methods.

2.7.3 PM Detectors and Modulators

Phase modulation is most frequently used for digital signals (PSK), because low BER can be obtained for relatively poor *S/N* ratios. The main drawback is the relatively broad spectrum produced by sudden phase changes. Many variations of phase modulation exist to improve the spectrum utilization. Important performance parameters are insertion loss, 180° phase difference between "high" and "low," and balanced or unbalanced operation requirement.

The PIN diode in Figure 2.45 is turned on or off by the digital bit stream, thereby changing the phase of the reflected signal appearing at RF_{out} by 180°. The insertion loss from RF_{in} to RF_{out} is about 6 dB; PIN diode reverse capacitance and lead inductance will produce some phase errors.

In Figure 2.46, the digital bit stream turns on either D_1 and D_3 or D_2 and D_4 in pairs, so that the RF and LO transformers are connected in phase or out of phase, depending on the polarity of the digital input signal. The insertion loss of this PSK modulator is about 2 dB, and the upper frequency limit is about 700 MHz.

High-frequency, narrowband phase modulators can be constructed by switching in additional $\lambda/2$ transmission line length when 180° of phase shift is required.

PSK demodulation requires similar receiver processing as FM demodulation: amplification, amplitude limiting, and detection. Since there is a 180° ambiguity in the recovered carrier phase in Figure 2.47, the data-out bit stream may be TRUE

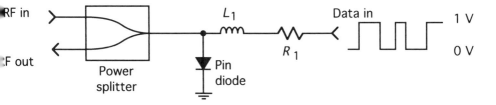

Figure 2.45 PSK modulator using a power splitter.

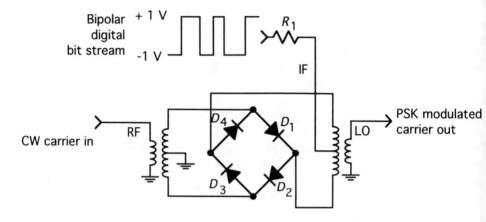

Figure 2.46 PSK modulator using a double–balanced mixer.

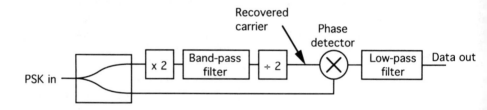

Figure 2.47 Coherent PSK detector.

or inverted. Differential encoding of the input data stream results in DPSK and avoids this difficulty.

$$b_n = a_n \oplus b_{n-1} \tag{2.16}$$

b_n = encoded message bit

a_n = input message bit

b_{n-1} = previous encoded bit

$\oplus$ = exclusive OR operation

After coherent PSK detection, the original message can be decoded by a similar process:

$$a_n = b_n \oplus b_{n-1} \tag{2.17}$$

b_n = received bit

a_n = decoded message bit

b_{n-1} = previous received bit

$\oplus$ = exclusive OR operation

DPSK demodulation can be considerably simplified by using the previous bit phase as the reference, as shown in Figure 2.48.

The transmitted bit stream must still be encoded according to (2.16), but the received data out corresponds to the original message, and (2.17) need not be applied when using the demodulation method in Figure 2.48. The delay τ is equal to the bit delay. The BER for DPSK is slightly worse than for fully coherent PSK demodulation because errors tend to occur in pairs.

A double-balanced mixer can be used as a phase detector, because the IF frequency response extends down to dc. Some mixers are specially designed to operate as phase detectors. The output is zero for the two input signals in quadrature (90° phase difference).

The carrier recovery for many complex signals can also be accomplished by implementing the Costas loop [15], which is particularly suitable for recovering quadrature multiplexed signals.

2.8 DIPLEXERS

The purpose of a diplexer network is to provide a constant resistive impedance at its input terminal while providing some frequency selectivity at the output terminal. A diplexer is really a special type of power splitter that directs the signal to different ports, depending on the frequency, while maintaining a constant resistive input impedance. This is especially desirable for mixer IF port termination, where a constant resistive termination improves mixer linearity (no signals or harmonics are reflected back to be remixed when a diplexer of the appropriate impedance is used).

Two simple diplexer types are shown in Figure 2.49 and Figure 2.50.

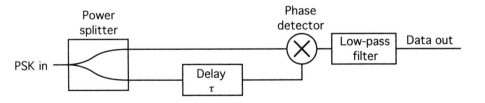

Figure 2.48 DPSK detector using previous bit phase comparison.

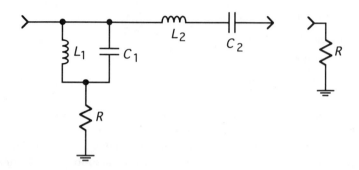

Figure 2.49 Bandpass diplexer #1.

$$Q = \frac{f_0}{BW_{3\,\text{dB}}}$$

$$L_2 = \frac{Q \times R}{2\,\pi f_0}$$

$$C_2 = \frac{1}{L_2(2\,\pi f_0)^2}$$

$$L_1 = \frac{R}{Q \times 2\,\pi f_0}$$

$$C_1 = \frac{1}{L_1(2\,\pi f_0)^2}$$

Q = relative bandwidth of through response

f_0 = center frequency of diplexer (Hz)

$BW_{3\text{dB}}$ = 3-dB bandwidth of desired through response (Hz)

R = terminating impedances of diplexer (Ω)

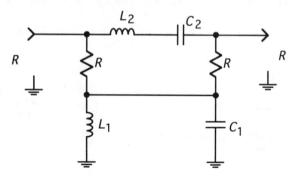

Figure 2.50 Bilateral bandpass diplexer.

$$Q = \frac{f_0}{BW_{3\,\text{dB}}}$$

$$L_2 = \frac{R\,Q}{2\,\pi f_0}$$

$$C_2 = \frac{1}{2\,\pi f_0 R\,Q}$$

$$L_1 = \frac{R}{2\,\pi f_0 Q}$$

$$C_1 = \frac{Q}{2\,\pi f_0 R}$$

Q = relative bandwidth of through response

f_0 = center frequency of diplexer (Hz)

BW_{3dB} = 3-dB bandwidth of desired through response (Hz)

R = terminating impedances of diplexer (Ω)

Diplexers can also be used to stabilize a potentially unstable amplifier, when the instability is outside the bandwidth of interest. At f_0, the diplexer network is essentially transparent, L_2 and C_2 are series resonant (short circuit), and L_1 and C_1 are parallel-resonant (open circuit). At frequencies above and below f_0, the input impedance is resistive and can effectively stabilize most cases of amplifier out-of-band instability.

At higher frequencies, where broadband 90° power splitters are practical, the arrangement in Figure 2.51 can be used to implement the diplexer function.

Any energy reflected from the filters appears at the bandstop output, while the through signal recombines at the bandpass output. The result is that the input is always matched, and signal energy is directed to one of the two outputs based on frequency. The bandwidth of the 90° hybrid must be wider than the bandwidth of the bandpass filters.

Diplexer networks of greater complexity and higher selectivity can be found in [16]. Low-pass/high-pass diplexers can be designed as singly terminated filters

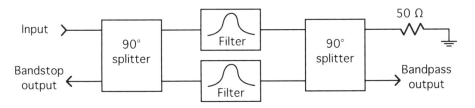

Figure 2.51 Diplexer using 90° hybrids and two identical filters.

or by computer optimization, making sure that the low-pass section starts with a series inductor, while the first component in the high-pass section is a series capacitor. Figure 2.52 shows this type of diplexer with 100-MHz corner frequency.

2.9 DIRECTIONAL COUPLERS

Directional couplers are most frequently used to sample the output of a power amplifier stage, with the purpose of controlling its forward power level. Other uses of directional couplers are for detecting antenna faults, measuring unknown impedances, and combining signals. Important directional coupler properties are low insertion loss and high directivity, which is a measure of the coupler's capability to detect power flow in only one direction. Operating bandwidth, coupling flatness, coupling accuracy, VSWR, and power-handling capability are also important in some applications.

$$\text{Coupling} \approx 20 \, \log(2\pi f C Z_0) \ (\text{dB})$$

$$\text{Coupling} < -15 \ \text{dB}$$

f = operating frequency (Hz)

Z_0 = characteristic impedance of $\lambda/4$ lines and terminating impedances (Ω)

C = coupling capacitance (F), $C < 0.18 \, / \, (2\pi f Z_0)$

Operation of the directional coupler shown in Figure 2.53 can be summarized by noting that power incident into port 1 is transferred to port 2 with little loss and is coupled into port 4 at the desired level of coupling. Port 3 is isolated. If power enters port 2 and leaves port 1, then port 3 is the coupled port and port 4

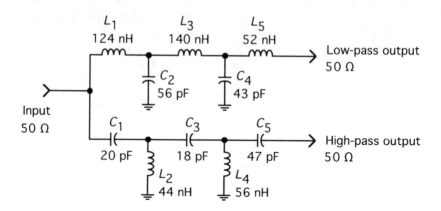

Figure 2.52 Low-pass/high-pass type of diplexer.

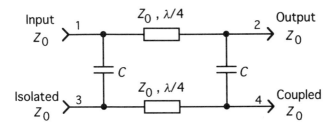

Figure 2.53 Generalized directional coupler.

is the isolated port. The coupling is relatively broadband; the bandwidth of operation is mainly limited by the desired isolation. If the coupled signal is to exceed the isolated signal by at least 20 dB (i.e., 20-dB directivity), the expected bandwidth will be about 12% for a 50-Ω coupler.

The two $\lambda/4$ transmission lines can be approximated by their lumped component equivalents, but for directional couplers the equivalent T-circuit shows better performance than the pi-circuit. The schematic diagram is shown in Figure 2.54, yielding about 10% bandwidth for 20-dB directivity.

$$L = \frac{Z_0}{2\pi f}$$

$$C = \frac{1}{2\pi f Z_0}$$

$$C_c \approx \frac{10^{(CF/20)}}{2\pi f Z_0}$$

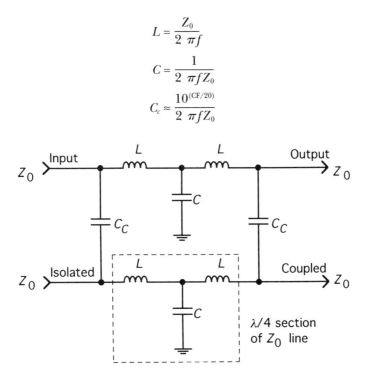

Figure 2.54 Lumped-element approximation of loosely coupled directional coupler.

f = operating frequency (Hz)

Z_0 = characteristic impedance of $\lambda/4$ lines and terminating impedances (Ω)

C_c = coupling capacitance (F), $C_c < 0.18/(2\pi f Z_0)$

CF = coupling factor (dB), CF < −15 dB

The topology in Figure 2.55 can be used for coupling tighter than 15 dB; 3-dB coupler design is shown:

Power incident into port 1 is coupled to port 2 at −3 dB, 0°; to port 4 at −3 dB, −90°; and port 3 is isolated, provided all ports are terminated in the characteristic impedance, Z_0. This device, therefore, also behaves as a 90° power splitter. The 90° phase difference is very broadband, but the power split is within 3 dB ±1 dB for approximately 20% bandwidth.

$$L = \frac{Z_0}{2\pi f}$$

$$C = \frac{1}{2\pi f Z_0}$$

The most familiar directional coupler consists of two $\lambda/4$ transverse electromagnetic mode (TEM) coupled lines, as shown in Figure 2.56, useful for bandwidths up to an octave. Note that for this configuration the coupled and isolated ports are reversed to what they were in the couplers in Figures 2.53, 2.54, and 2.55.

All ports must be terminated in the system characteristic impedance. The coupler in Figure 2.56 works best for TEM mode transmission lines, such as coaxial lines or striplines. It does not work very well in microstrip implementation, because

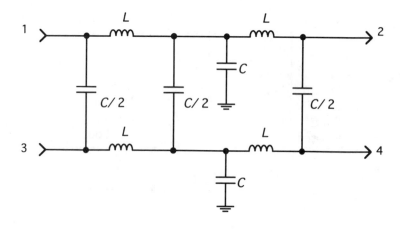

Figure 2.55 Narrowband lumped-element 3-dB coupler.

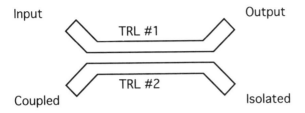

Figure 2.56 Coupled $\lambda/4$ transmission line directional coupler.

the even- and odd-mode velocity factors are different, resulting in inadequate signal cancellation at the isolated port, that is, the coupler loses directivity. At the maximum coupling of 3 dB, the coupler properties are the same as a 90° power splitter. The bandwidth of this type of coupler can be extended by using more $\lambda/4$ sections [17].

$$CF = 20 \log \left(\frac{\dfrac{Z_{0e}}{Z_{0o}} - 1}{\dfrac{Z_{0e}}{Z_{0o}} + 1} \right) \qquad (2.18)$$

$$Z_0 = \sqrt{Z_{0e} Z_{0o}} \qquad (2.19)$$

CF = coupling factor (dB)

Z_{0e} = even-mode characteristic impedance (Ω)

Z_{0o} = odd-mode characteristic impedance (Ω), $Z_{0o} < Z_{0e}$

Z_0 = system impedance (Ω)

A directional coupler is frequently used to sample the output signal of a transmitter and the coupled port is connected to a diode detector. In that case, transmission line 2 characteristic impedance is designed to be higher than 50 Ω (narrower trace), to develop greater voltage across the detector diode for given power coupling.

Power amplifiers designed to operate over a wide range of output powers impose some interesting system constraints on the directional coupler that samples the output power: The coupling factor must be sufficiently high to reliably sample the amplifier output at the lowest power setting, but such a tight coupling causes harmonics generated by the detector diodes to couple back into the output at the highest power setting. Therefore, the directional coupler should be followed by a low-pass filter to attenuate those harmonics. The obvious disadvantage of such an arrangement is that the power-leveling loop can no longer compensate for the low-pass filter loss.

The power-sampling directional coupler should not be placed right at the output of a high-power stage, because its output is usually rich in harmonics, which may be preferentially coupled to the detector (depending on the directional coupler implementation).

One of the simplest and most generic lumped-element directional coupler topologies consists of two mutually coupled coils and a capacitor, as shown in Figure 2.57.

Lines 1-2 and 3-4 are both capacitively and inductively coupled to each other. Particular values of L_1, L_2, C_{12}, and k are required for this circuit to behave as a directional coupler; in particular, $\omega L_1 \ll Z_0$, $\omega L_2 \ll Z_0$.

$$ CF \approx 20 \log \left(\frac{Z_0}{Z_0 + \dfrac{1}{2 \pi f C_{12}}} \right) \qquad (2.20) $$

CF = coupling factor (dB), CF < 0

Z_0 = characteristic impedance, terminating impedance at all ports (Ω)

f = desired frequency of operation (Hz)

Equation (2.20) shows that the coupling will increase with frequency. L_1, L_2, and k depend on each other in a complex manner but can be easily determined by computer optimization. The circuit in Figure 2.58 shows a 50-Ω, −26-dB coupler at 160 MHz. The bandwidth of this circuit is determined by the coupling accuracy desired and can be well over an octave if a slight increase of coupling with frequency can be tolerated.

Such a circuit can also be realized by two microstrip lines of equal width in close proximity, as shown in Figure 2.59.

$$ C_{12} = \frac{x}{2\,c} \left(\frac{1}{v_{po} Z_{0o}} - \frac{1}{v_{pe} Z_{0e}} \right) $$

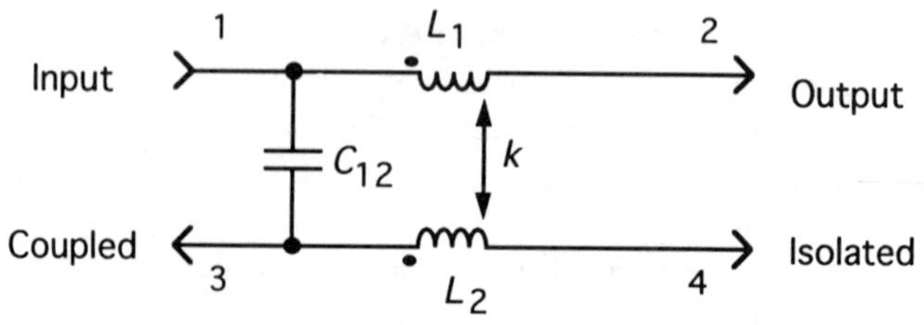

Figure 2.57 Generic lumped-element directional coupler.

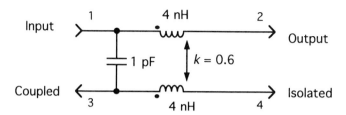

Figure 2.58 A 26-dB VHF directional coupler.

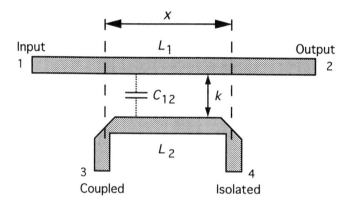

Figure 2.59 Printed directional coupler much shorter than $\lambda/4$.

x = length of coupling section (m)

C_{12} = equivalent coupling capacitance (F), which determines coupling factor

c = speed of light, 2.9979×10^8 (m/s)

v_{po} = odd-mode velocity factor, $0 < v_{po} < 1$

v_{pe} = even-mode velocity factor, $0 < v_{pe} < 1$

Z_{0o} = odd-mode characteristic impedance (Ω)

Z_{0e} = even-mode characteristic impedance (Ω)

This circuit is much shorter than a quarter-wavelength yet operates over a wide bandwidth, with the disadvantage that the coupling increases with frequency. The line widths, coupling length, and spacing must be designed properly for correct operation. In contrast to the full $\lambda/4$ coupler in Figure 2.56, this circuit can be realized in microstrip.

When such a coupler is implemented on a high ϵ_r circuit board, such as G10 or alumina, the equivalent C_{12} is usually much higher than required for the corresponding k. To compensate for this effect, the coupling line is made narrower, which has the effect of decreasing C_{12} and increasing k. The example shown in Figure 2.60 illustrates this: the main line requires 50 Ω terminations, while the coupled line requires 100 Ω terminating impedances. The higher characteristic impedance of the coupled line has the added benefit of increasing the RF voltage for a given amount of coupled power, thus improving the sensitivity of the diode detector used for sensing the power, as previously mentioned.

This coupler is about $\lambda/25$ long at 125 MHz and has 30 dB coupling and 20 dB directivity. The coupling increases with frequency. Electrical parameters for this line, implemented in microstrip on G-10 material, are:

$Z_{0e}{}^a = 54.357\ \Omega$

$v_{pe}{}^a = 0.573$

$Z_{0o}{}^a = 44.29\ \Omega$

$v_{po}{}^a = 0.599$

$Z_{0e}{}^b = 95.7\ \Omega$

$v_{pe}{}^b = 0.584$

$Z_{0o}{}^b = 68.43\ \Omega$

$v_{po}{}^b = 0.622$

The electrical equivalent circuit for the configuration in Figure 2.60 is shown in Figure 2.61.

Z_{11} and Z_{22} are unbalanced transmission lines to ground, and Z_{12} is a balanced line.

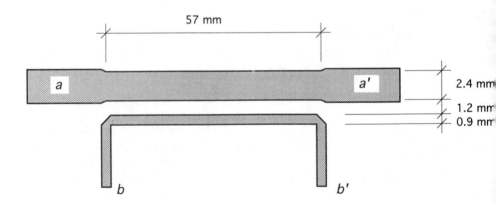

Figure 2.60 A 30-dB 125-MHz microstrip directional coupler.

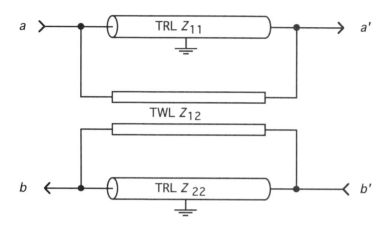

Figure 2.61 Equivalent circuit of two asymmetrical coupled microstrip lines.

$$Z_{11} = Z_{0e}^a$$
$$Z_{22} = Z_{0e}^b$$

$$Z_{12} = \cfrac{2}{\sqrt{\left(\dfrac{v_{po}^a}{Z_{0o}^a} - \dfrac{v_{pe}^a}{Z_{0e}^a}\right)\left(\dfrac{1}{v_{po}^a Z_{0o}^a} - \dfrac{1}{v_{pe}^a Z_{0e}^a}\right)}}$$

$$v_{11} = v_{pe}^a$$
$$v_{22} = v_{pe}^b$$

$$v_{12} = \sqrt{\cfrac{\dfrac{v_{po}^a}{Z_{0o}^a} - \dfrac{v_{pe}^a}{Z_{0e}^a}}{\dfrac{1}{v_{po}^a Z_{0o}^a} - \dfrac{1}{v_{pe}^a Z_{0e}^a}}}$$

Z_{0e}^a = even-mode impedance when a is the reference conductor
v_{pe}^a = even-mode velocity factor when a is the reference conductor
Z_{0o}^a = odd-mode impedance when a is the reference conductor
v_{po}^a = odd-mode velocity factor when a is the reference conductor
Z_{0e}^b = even-mode impedance when b is the reference conductor
v_{pe}^b = even-mode velocity factor when b is the reference conductor
Z_{0o}^b = odd-mode impedance when b is the reference conductor

v_{po}^b = odd-mode velocity factor when b is the reference conductor

Z_{11} = characteristic impedance of unbalanced line a in equivalent circuit

Z_{22} = characteristic impedance of unbalanced line b in equivalent circuit

Z_{12} = characteristic impedance of balanced line in equivalent circuit

v_{11} = velocity factor of line Z_{11}

v_{22} = velocity factor of line Z_{22}

v_{12} = velocity factor of line Z_{12}

The lengths of all the above transmission lines are the same and are equal to the physical length of the coupling.

A number of computer programs are available to compute the mode impedances and velocity factors from the physical geometry of asymmetrical coupled microstrip lines: RFLaplace™ from ingSOFT [18], Structure Simulator from Hewlett-Packard [19], and Maxwell® Spicelink from Ansoft [20].

The circuit shown in Figure 2.62 is yet another directional coupler topology, suitable for multioctave operation.

Power fed into port 1 emerges at port 2 and is sampled at port 3. Power fed into port 2 emerges at port 1, with no power coupled to port 3. Terminating impedance at all ports is $R\Omega$ to ground.

$$CF = 20 \log\left(\frac{R}{R_1 + R}\right)$$

$$IL = -20 \log\left(\frac{R}{R_2 + R}\right)$$

$$R = \sqrt{R_1 R_2}$$

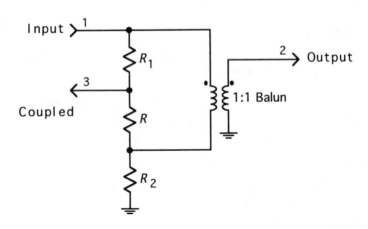

Figure 2.62 Broadband directional coupler using a balun.

CF = coupling factor, coupled power relative to input power (dB), CF < 0

IL = insertion loss from port 1 to port 2 (dB), IL > 0

R = system impedance (Ω)

The three quantities—coupling factor, insertion loss, and system impedance—are not independent; once a coupling factor in a 50-Ω system is selected, the insertion loss becomes fixed. The directional coupler in Figure 2.62 can be used as a broadband unequal power splitter with isolation. For example, the circuit in Figure 2.63 is a 50-Ω 10-dB / 3.3 dB power splitter with theoretical 60-dB isolation between ports 2 and 3.

The performance and bandwidth of this circuit are limited only by the balun and stray parasitics.

A very small, low-frequency directional coupler can be conveniently constructed on a two-hole ferrite core using multifilar windings, as shown in Figure 2.64. The primary typically contains very few turns, while the ratio of secondary to

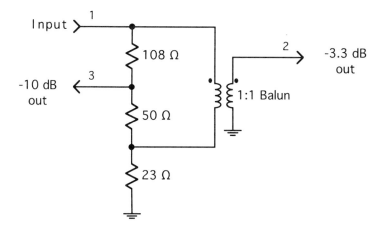

Figure 2.63 A 3.3-dB/10-dB unequal power splitter with isolation.

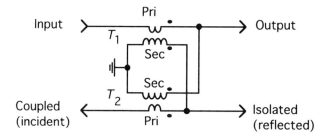

Figure 2.64 Directional coupler using two transformers.

primary turns determines the coupling. The upper frequency range of this device rarely exceeds 600 MHz.

$$C \approx 20 \log n$$

C = coupling (dB)

n = secondary to primary turns ratio (linear)

A turns ratio of 1.414 results in a 3-dB, 180° power splitter when the unused (isolated) port is terminated in 100 Ω. This is in contrast to other reactive couplers in this section, which have a 90° phase difference between the output and coupled ports.

Any directional coupler can be used as an unequal power splitter, or *tap* in cable communication terminology.

2.10 FILTERS

Filters are the most fundamental building blocks for achieving frequency selection, transmitting certain frequencies without attenuation while rejecting other frequencies. Impedance-matching networks are also by necessity filter networks as a result of the Bode-Fano relationship, which is described in Section 2.12. The basic tradeoff in filter design is low *VSWR* in the passband and sufficient attenuation in the stopband. The majority of filters achieve frequency selection by reflection; a small class of filters called diplexers or absorptive filters achieve attenuation by absorbing the incoming energy while presenting a good impedance match in both passband and stopband. The amplitude and phase responses of an *LC* filter's transfer function are not independent but are related through the Hilbert transform. In practical terms, filters with very steep transition bands have the largest group delay, and, conversely, flat group delay requires very gentle transition from passband to stopband. SAW resonator filters and some digital filter implementations avoid this difficulty and can shape the amplitude and phase responses independently.

This section assumes that the filters will be realized with ideal components and is therefore only a first step in obtaining the desired filter. A realistic filter design must account for component parasitics (see Section 3.2), circuit board pad capacitances, and trace inductances. The parasitic components and losses in practical filters all conspire to lower the corner frequency, decrease bandwidth, and increase insertion loss.

After designing the ideal filter by the methods in this section, decide on a layout, include pad capacitances, model traces as transmission line sections, add component parasitics, use computer optimization to adjust the filter component values, and recover the desired performance. The key concept is to include the circuit layout in the modeling process; you cannot accurately model RF circuits without taking the layout into account.

2.10.1 LC Filters

The two most common lowpass filters were selected for this section: Butterworth and Chebyshev responses. Both filters share the same topology shown in Figure 2.65 and Figure 2.66; Butterworth filters have lower insertion loss for given component Q, while Chebyshev filters have steeper selectivity slope.

Butterworth filter design equations:

$$C_{k,\text{odd}} = \frac{g_k}{2\,\pi f R} \tag{2.21}$$

$$L_{k,\text{even}} = \frac{g_k R}{2\,\pi f} \tag{2.22}$$

$$g_k = 2\,\sin\left(\frac{(2\,k-1)\,\pi}{2n}\right) \tag{2.23}$$

f = 3-dB corner frequency (Hz)

R = source and load resistance (both the same) (Ω)

n = number of components in filter

k = component number, odd for shunt C, even for series L

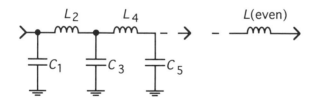

Figure 2.65 Butterworth low–pass filter topology.

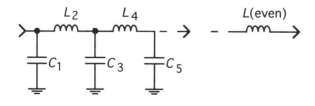

Figure 2.66 Chebyshev low–pass filter topology.

The dual topology, which starts with a series L rather than shunt C, is less preferred, because it cannot absorb the inevitable shunt capacitance at the input and output; n should be odd for the same reason.

Chebyshev filter design equations:

$$C_{k,\text{odd}} = \frac{g_k}{2\,\pi f R} \tag{2.24}$$

$$L_{k,\text{even}} = \frac{g_k R}{2\,\pi f} \tag{2.25}$$

$$g_1 = \frac{2a_1}{\sinh\left\{\dfrac{\ln\left[\coth\left(\dfrac{L_{Ar}}{17.37}\right)\right]}{2n}\right\}} \tag{2.26}$$

$$g_k = \frac{4a_{k-1}a_k}{b_{k-1}g_{k-1}}, \qquad k = 2, 3, \ldots, n \tag{2.27}$$

$$a_k = \sin\left(\frac{(2\,k-1)\,\pi}{2n}\right), \qquad k = 1, 2, \ldots, n \tag{2.28}$$

$$b_k = \sinh^2\left\{\frac{\ln\left[\coth\left(\dfrac{L_{Ar}}{17.37}\right)\right]}{2n}\right\} + \sin^2\left(\frac{k\,\pi}{n}\right), \qquad k = 1, 2, \ldots, n \tag{2.29}$$

$$R_L = R \text{ for } n \text{ odd,} \tag{2.30}$$

$$R_L = R \tanh^2\left\{\frac{\ln\left[\coth\left(\dfrac{L_{Ar}}{17.37}\right)\right]}{4}\right\} \text{ for } n \text{ even} \tag{2.31}$$

f = ripple (not 3 dB) corner frequency (Hz)

R = source resistance (Ω)

L_{Ar} = passband ripple (dB)

n = number of components in filter

R_L = required load resistance (Ω)

k = component number, odd for shunt C, even for series L

Many references use the 3-dB bandwidth rather than the ripple bandwidth in tabulating Chebyshev filter component values. The two bandwidths are related by:

$$\frac{f_{3\,dB}}{f_{ripple}} = \cosh\left[\frac{1}{n}\cosh^{-1}\left(\frac{1}{\sqrt{10^{L_{Ar}/10}-1}}\right)\right] \tag{2.32}$$

Since the passband ripple of Chebyshev filters is due entirely to mismatch loss, the in-band return loss and passband ripple are related by (4.33), which can be rewritten as:

$$10^{-L_{Ar}/10} + 10^{-RL/10} = 1$$

Highpass, bandpass, and bandstop filters can be derived from low-pass proto-types using the transformations in Section 4.4. The design of bandpass filters by a low-pass to bandpass transformation usually results in component values that are difficult to realize. Another technique, based on coupled resonators, is particularly convenient if properly understood. It consists of identical series or shunt resonators coupled to each other through impedance inverters. What makes this technique particularly useful is that any type of resonator can be used: *LC*, transmission line, or crystal resonators. Impedance inverters can be approximated by a number of narrowband circuits to implement a large variety of useful circuits. The design steps for coupled resonator bandpass filters can be summarized by the following five-step procedure.

1. Decide on a resonator to be used. The only constraint required is that its impedance (*L* to *C* ratio, or susceptance/reactance slope) properties be known.
2. Calculate the *k* and *q* coupling coefficients appropriate for the design (Butter-worth, Chebyshev, etc.). This can be done by using g_k coefficients from (2.23) and (2.27). The variable *n* in those equations becomes the number of resona-tors used.

$$k_{j,j+1} = \frac{1}{\sqrt{g_j g_{j+1}}} \tag{2.33}$$

$$q_0 = g_1, \qquad q_n = g_n \tag{2.34}$$

The appropriate *k* and *q* coefficients for other types of filters are also available in tables. The use of the 3-dB or the ripple frequency for Chebyshev filters must be clearly differentiated, because the *k* and *q* coefficients are different depending on which reference bandwidth we wish to use. The two sets of coefficients are related by

$$q_{3\,dB} = q_{ripple}\frac{f_{3\,dB}}{f_{ripple}} \tag{2.35}$$

$$k_{3\,dB} = k_{ripple}\frac{f_{ripple}}{f_{3\,dB}} \tag{2.36}$$

$q_{3\,dB}$ = input/output coupling coefficient when 3-dB bandwidth is used as reference for Chebyshev filters

q_{ripple} = input/output coupling coefficient when ripple bandwidth is used as reference for Chebyshev filters, derived from (2.26) through (2.34)

$k_{3\,dB}$ = coupling coefficient when 3-dB bandwidth is used as reference for Chebyshev filters

k_{ripple} = coupling coefficient when ripple bandwidth is used as reference for Chebyshev filters, derived from (2.26) through (2.34)

$f_{3\,dB}$ = 3-dB bandwidth reference for Chebyshev filters, related to f_{ripple} by (2.32)

f_{ripple} = ripple bandwidth reference for Chebyshev filters, related to $f_{3\,dB}$ by (2.32)

3. Decide on the appropriate impedance inverter topology, whose negative components can be easily absorbed by adjacent resonators. Calculate impedance inverter component values. Figure 2.67 shows some of the possibilities. Another set of inverters to choose from can be obtained from Figure 2.67 by using $-C_i$ wherever L_i is required and $-L_i$ whenever C_i is shown. A $\lambda/4$ length of transmission line with its characteristic impedance equal to K, realized either as distributed or its lumped-component equivalent, will also serve very well as an impedance inverter circuit.

   The inverter K value (and from it the equivalent coupling inductance or capacitance value) can be obtained from the following expressions.

$$K_{j,j+1}^{S} = \frac{k_{j,j+1}BW_{3\,dB}}{f_0}\sqrt{X_j X_{j+1}} \tag{2.37}$$

$$K_{j,j+1}^{P} = \frac{f_0}{k_{j,j+1}BW_{3\,dB}}\sqrt{X_j X_{j+1}} \tag{2.38}$$

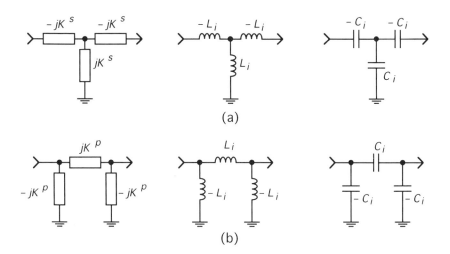

Figure 2.67 Impedance inverter approximations: (a) suitable for coupling series resonators, $K^S = \omega_0 L_i = 1/(\omega_0 C_i)$; (b) suitable for coupling parallel resonators, $K^p = \omega_0 L_i = 1/(\omega_0 C_i)$.

$K^s_{j,j+1}$ = impedance inverter reactance for coupling series resonators,

$\quad = \omega_0 L_i = 1/(\omega_0 C_i)$

$K^p_{j,j+1}$ = impedance inverter reactance for coupling parallel resonators,

$\quad = \omega_0 L_i = 1/(\omega_0 C_i)$

$k_{j,j+1}$ = normalized coupling coefficient between resonators,

$\quad$ computed from (2.33)

$BW_{3\,dB}$ = 3-dB bandwidth of desired filter (Hz)

f_0 = center frequency of desired filter (Hz)

ω_0 = center frequency of desired filter (rad/s)

X_j = resonator reactance slope at resonance,

$\quad = \omega_0 L = 1/(\omega_0 C)$, where L and C are the

$\quad$ resonator equivalent component values

L_i, C_i = impedance inverter component values

L, C = resonator equivalent circuit values

4. Calculate required terminating impedances for the filter.

$$R^S = \frac{X_1 BW_{3\,dB}}{q_1 f_0} \text{ for series resonator topology} \qquad (2.39)$$

$$R^P = \frac{X_1 q_1 f_0}{BW_{3\,\text{dB}}} \text{ for parallel resonator topology} \qquad (2.40)$$

5. The filter implementation will introduce parasitic components such as coil losses, circuit trace inductances, and node capacitances, which are difficult to include in the synthesis procedure and are best handled by computer optimization of an accurate model of the proposed filter layout.

The series resonator design procedure is illustrated by the example in Figure 2.30, while the parallel resonator case is expanded in Subsection 2.10.2. Additional design information can also be found in [21].

Chebyshev low-pass filters of even order require different source and load impedances in their realization; this is not so for Chebyshev bandpass filters. Regardless of filter order, the source and load impedances are always equal for bandpass filters realized using the coupled resonator technique.

Subsection 2.10.2 expands on the five-step design process described above, and presents additional information on modeling and empirical adjustment of high Q, coupled-resonator bandpass filters.

2.10.2 Coupled Resonator Filters

Many of us have at one time or another been impressed by large base station filters, reminiscent of small engine blocks, whose impressive performance places them in a class by themselves. Yet such filters are quite easy to design and simulate; the main difficulties associated with these filters are mechanical. The main reason for using coupled resonator filters is that very high component Q's (and therefore low insertion loss and high selectivity) can be achieved. Conventional LC filters are limited by coil Q of about 150, while the Q of a coaxial or helical resonator is limited only by its physical size. Resonator Q's in the high hundreds or even thousands are achievable in practical structures. Coupled resonator filters are usually constructed using high- Q, $\lambda/4$ resonators coupled to each other by means of openings or apertures in the walls separating the resonators from each other. Such a structure produces a narrow bandwidth filter, which can be modeled by a cascade of parallel LCR sections and impedance inverters. A three-resonator filter example is shown in Figure 2.68.

All the inductors, capacitors, and resistors are equal; input and output coupling is modeled using ideal transformers; and interresonator coupling is modeled by impedance inverters. There are several degrees of freedom in this schematic; no unique combination of components will result in a specific frequency response. No measurement can be performed at the filter terminals that will tell us what the values of L, C, K, and R are. This is really quite an amazing fact; instead of presenting a design difficulty, it opens up almost infinite possibilities: We can choose helical

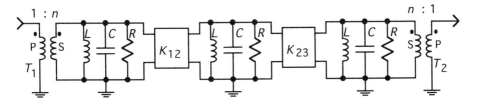

Figure 2.68 Coupled parallel resonator filter of three sections.

or coaxial resonators, ceramic blocks, or any resonant structure to realize such filters! The only real requirement is that the unloaded Q be high enough to ensure reasonable insertion loss through the filter. The four parameters that must be known before we can start assigning component values are center frequency of filter f_0, unloaded Q of each resonator, filter bandwidth Δf, and desired coupling coefficients (i.e., Butterworth, Chebyshev shape), obtained from Subsection 2.10.1 and (2.33) and (2.34).

A convenient set of component values for the equivalent circuit in Figure 2.68 can be generated as follows. Note that these are the same basic equations as developed in previous sections, except that they have been simplified by our convenient choice of C and L.

$$L = \frac{1}{\omega_0} \tag{2.41}$$

$$C = \frac{1}{\omega_0} \tag{2.42}$$

$$R = Q \tag{2.43}$$

$$K_{ij} = \frac{f_0}{\Delta f k_{ij}} \tag{2.44}$$

$$n = \sqrt{\frac{f_0 \, q_1}{R_S \, \Delta f}} \tag{2.45}$$

L = parallel inductance in equivalent circuit (H)

C = parallel capacitance in equivalent circuit (F)

R = parallel resistance in equivalent circuit (Ω)

Q = unloaded quality factor of resonator

K_{ij} = impedance inverter characteristic impedance (Ω)

Δf = filter bandwidth (Hz)

k_{ij} = normalized coupling coefficients obtained from Subsection 2.10.1

n = turns ratio of transformer in equivalent circuit

ω_0 = center frequency of filter (rad/s)

f_0 = center frequency of filter = $\omega_0/(2\pi)$ (Hz)

q_1 = normalized input/output coupling coefficient obtained from
  Subsection 2.10.1

R_S = source resistance (Ω)

To illustrate the procedure for generating an equivalent circuit, assume that we wish to model a five-resonator, 8-MHz-wide filter operating at 810 MHz, using $\lambda/4$ resonators with unloaded Q of 1800. It is to be a Chebyshev filter with k_{12} = k_{45} = 0.79745, k_{23} = k_{34} = 0.6077, and q_1 = q_5 = 1.1468. Then, from (2.41) through (2.45):

$L = 0.1965$ nH

$C = 196.5$ pF

$R = 1800\ \Omega$

$K_{12} = 126.97\ \Omega$

$K_{45} = 126.97\ \Omega$

$K_{23} = 166.61\ \Omega$

$K_{34} = 166.61\ \Omega$

$n = 1.5239$

The equivalent circuit is shown in Figure 2.69.

The frequency response of this circuit can be obtained by conventional circuit analysis programs and is shown in Figure 2.70.

The equivalent circuit for the desired filter uses no information about the physical filter, except for the unloaded Q. Why would such a general filter description be useful? This question can be answered in four parts:

The implementation of high-performance coupled resonator filters relies on empirical adjustment of coupling apertures and input/output coupling. The electri-

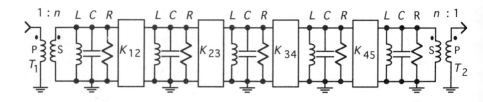

Figure 2.69 Five-resonator Chebyshev filter equivalent circuit.

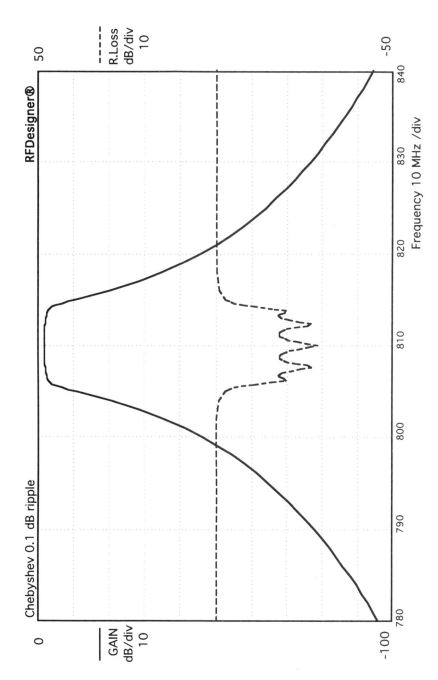

Figure 2.70 Five-resonator Chebyshev filter frequency response.

cal equivalent circuit allows us to determine the required coupling parameters by computer simulation, including the effects of finite Q. The two coupling parameters are $\Delta 3dB$ and Δf_p, as defined by Zverev [22].

$\Delta 3dB$ = 3-dB bandwidth when lightly probing the first resonator, with second resonator shorted

Δf_p = frequency difference among frequency peaks at the nth resonator, when the $(n + 1)^{th}$ resonator is shorted to ground

These two measured parameters are related to the equivalent circuit defined by (2.41) to (2.45), when $Q = \infty$ by:

$$K_{ij} = \frac{f_0}{\Delta f_p} \tag{2.46}$$

$$n = \sqrt{\frac{f_0}{R_S\,\Delta 3dB}} \tag{2.47}$$

Thus, $\Delta 3dB$ correlates to n, and Δf_p correlates to K_{ij}. In the general case when $Q \neq \infty$, these equations cannot be used, and $\Delta 3dB$ and Δf_p have to be obtained by computer simulation from the equivalent circuit.

Getting away from the mathematical details for a moment, note that two sets of numbers set the filter type and bandwidth. The usual parameters are q and k, which we have transformed to n and K. Once the equivalent circuit component values have been assigned, we can obtain all the relevant parameters that would normally be measured. For example, the circuit in Figure 2.71 can be used to obtain $\Delta 3dB$ for the simulated filter, by shorting out the second resonator ($R = 0$) and probing the first resonator with a high-impedance probe:

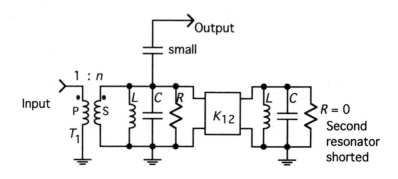

Figure 2.71 Determining required $\Delta 3dB$ from equivalent circuit by computer simulation.

The input coupling in the physical filter would then be empirically adjusted to obtain the required Δ3dB determined by simulation from the equivalent circuit. This method works for any type of input coupling: tap, coupling loop, or capacitive probe.

For our example in Figure 2.69, the required Δ3dB = 7.416 MHz. Note that in the theoretical case when $Q = \infty$, the required Δ3dB would be 6.976. The usual assumption of infinite Q, with its associated errors, need not be made, if we have access to the equivalent circuit and computer simulation.

Δf_p can be simulated or measured in a similar manner, shorting out the third resonator and lightly sampling the energy in the first resonator, as shown in Figure 2.72.

Two peaks in the frequency response will be obtained; their frequency difference will give Δf_p. This frequency difference is usually difficult to measure accurately, because the peaks are not very sharp. Our equivalent circuit allows us to devise other, more easily measured characteristics, such as the notch width between the two peaks at the 10-dB down points. Whatever parameter we decide to track, the aperture between the first and second resonators in the physical filter is adjusted until the desired Δf_p is obtained in the physical filter, corresponding to the Δf_p value obtained by computer simulation. The procedure is repeated for all apertures in the filter. The second reason the equivalent circuit is useful is its ability to handle resonators with unequal Q's. The appropriate R value is entered in the equivalent circuit, and simulated measurements can be performed without further difficulty.

The third consideration is that the transformer turns ratio and impedance inverter characteristic impedance can be related to physical dimensions of the filter. The mechanical tolerances can then be correlated to electrical tolerances, and Monte-Carlo simulation can be performed to ascertain the effect of those mechanical tolerances, as well as to estimate tuning uncertainties. Plating variations will contribute to R, internal apertures to K_{ij}, input/output coupling to n, and tuning uncertainty to ω_0. Their relative contributions are best determined by measurement.

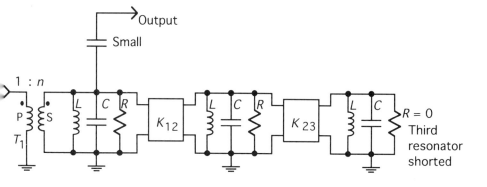

Figure 2.72 Determining required Δf_p from equivalent circuit by computer simulation.

For example, if we deliberately increase an aperture size and evaluate its effect of Δf_p and hence on K_{ij}, we can then relate the mechanical tolerance to tolerance on K_{ij} to be used in the model. Tuning uncertainty can be quantified by assigning a distribution to the center frequency (either inductance or capacitance can be varied to simulate tuning uncertainty). If peak-dip tuning is used, the amplitude resolution of the tuning indicator can serve to estimate the frequency accuracy of the desired peak or dip, based on their frequency responses from the equivalent circuit. Finally, the equivalent circuit can also be used to estimate the effect of temperature drift, usually the most challenging design problem of coupled resonator filters. The equivalent circuit can relate the frequency shift to insertion loss degradation at filter band edges. Once the required equivalent circuit is obtained at a given frequency, a set of convenient transformations can be used to arrive at different, related designs.

If we have a design at one frequency, f_1, and wish to obtain the identical bandwidth, return loss, and insertion loss at a different frequency, f_2, the following transformations in the equivalent circuit need to take place:

$$L_2 = C_2 = \frac{1}{\omega_2}$$

$$K_{ij,2} = K_{ij,1}\frac{f_2}{f_1}$$

$$n_2 = n_1\sqrt{\frac{f_2}{f_1}}$$

$$R_2 = R_1\frac{f_2}{f_1}$$

If we have a design of certain bandwidth Δf_1 and wish to obtain a different bandwidth Δf_2 at the same center frequency, with the same return loss and insertion loss, the following transformations are appropriate:

$$K_{ij,2} = K_{ij,1}\frac{\Delta f_1}{\Delta f_2}$$

$$n_2 = n_1\sqrt{\frac{\Delta f_1}{\Delta f_2}}$$

$$R_2 = R_1\frac{\Delta f_1}{\Delta f_2}$$

When measuring Δf_p or $\Delta 3dB$, the measuring equipment is operating in an environment where the impedances of the test item change a great deal. Care must be taken to avoid frequency pulling or amplitude ripple in the signal generator used for the measurement. At least 10 dB of attenuation should be used between the signal generator and the test item, illustrated in Figure 2.73.

A convenient method of detuning, or shorting, the other resonators must also be available for accurate measurements, although the equivalent circuit can take imperfect detuning into account, as long as we know the frequency of the detuned resonators. The coupling coefficients, which determine the filter shape, can be obtained from the low-pass filter prototypes introduced in Subsection 2.10.1, by setting the terminating impedance at 1 Ω and corner frequency at 1 rad/sec.

Again, be aware that the coupling coefficients are different, depending on the choice of 3 dB or ripple bandwidth for Chebyshev filters, as specified by (2.35) and (2.36).

$$q_1 = q_n = g_1$$

$$k_{ij} = \frac{1}{\sqrt{g_i g_j}}$$

q_1 = normalized input coupling coefficient

q_n = normalizing output coupling coefficient

k_{ij} = normalized coupling coefficient between resonators i and j

g_1 = first normalized prototype component value

g_i = normalized i^{th} prototype component value

g_j = normalized j^{th} prototype component value

$j = i + 1$

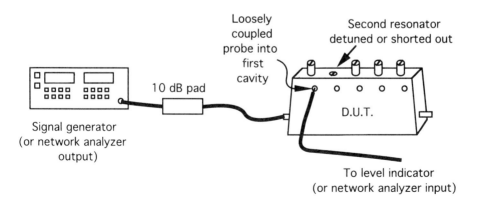

Figure 2.73 Using attenuator to minimize Δf_p and $\Delta 3dB$ measurement error.

The k and q values, as well as the n and K values, can also be obtained indirectly by computer optimization techniques. Simply decide on the filter performance goals, set n and K for optimization, and let the computer converge on the desired set of values.

Coupled resonator filters with relative 3-dB bandwidths of 0.5% to 15% are feasible; aperture coupling can be used up to about 3% relative bandwidth, after which direct coupling should be used to achieve wider bandwidths.

Sections of 50-Ω transmission line can be added at the input and output of the equivalent circuit to obtain the required delays, if phase information is important in your model.

2.10.3 Narrow-Notch Filters

A technique for constructing very narrow-notch filters is shown in Figure 2.74; a 50-kHz-wide notch at UHF is easily achievable. The connections between the circulator, T-connector, and cavity must be kept as short as possible.

The notch bandwidth that can be obtained is much narrower than using the cavity alone. This filter has the additional interesting property that its input impedance at the notch frequency is resistive; it does not achieve selectivity by reflection and impedance mismatch. Using a circulator results in the lowest insertion loss; if this is not required, a simple power splitter will work just as well. The principle of operation relies on the fact that when the impedance presented at circulator port 2 is equal to 50 Ω, then there is no output at port 3. A reasonable question might be how can we get a 50-Ω resistive impedance from the circuit connected to port 2, noting that it only contains reactive components?

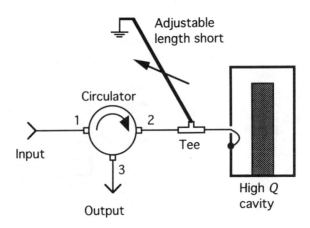

Figure 2.74 Narrowband absorptive notch filter.

The required resistance comes from the cavity's unloaded Q, which appears as a very high parallel resistance in the circuit, which is transformed to 50 Ω by the shunt tunable line and high Q cavity tuning. Since the impedance transformation is so extreme, the bandwidth over which port 2 sees a 50-Ω load is very narrow, and so the desired objective has been achieved. Both the cavity and the shunt line must be carefully tuned and isolated from mechanical and thermal shocks to achieve a stable notch. Other configurations of this basic topology are also possible [23].

Looking at the circuit in Figure 2.74 from an energy conservation point of view, we observe that all energy fed into the input (circulator port 1) must be dissipated in the cavity at the notch frequency. Since the resonator's equivalent parallel resistance is high—see (4.63), the RF voltage inside the cavity can be quite high; in fact, this is a similar principle that N. Tesla used in his famous coil experiments. Therefore, the setup in Figure 2.74 can also be used to generate very high RF voltages by employing a cavity of the required high quality factor.

2.11 FREQUENCY MULTIPLIERS

Frequency multipliers are used in situations where an oscillator at the high fundamental frequency is limited in some way: there is insufficient device gain, the resonator is impractical, phase noise is too high, or wideband frequency modulation is desired.

Basic Fourier theory [14] can indicate which devices or methods may be suitable for frequency multiplication. For example, a square wave has no even harmonics, and the third harmonic is 7 dB down from the peak input signal. Therefore, a limiter will perform poorly as a frequency doubler, and the best conversion loss that can be expected for the third harmonic is 7 dB.

Half-wave and full-wave rectified signals, on the other hand, have no odd harmonics. Conversion loss from peak input for the second harmonic is 13.5 dB for half-wave and 7.5 dB for full-wave rectifier. An active class-C frequency multiplier can therefore be expected to operate slightly more efficiently for even harmonics than for odd harmonics, unless driven so hard that limiting occurs.

Whatever circuit is used for frequency multiplication, the output will be rich in undesired harmonics; for that reason, frequency multipliers are inherently narrowband devices, because the desired output harmonic must be selected by a bandpass filter.

A double-balanced mixer together with a power splitter can be used as a frequency doubler, as shown in Figure 2.75(a). The conversion loss for the second harmonic is about 18 dB, with all other harmonics (including fundamental) more than 12 dB down from the second harmonic signal.

Any full-wave rectifier circuit, such as the one shown in Figure 2.75(b), achieves about 15-dB conversion loss for the second harmonic, with all other harmonics (including fundamental) more than 10 dB down from the second harmonic signal.

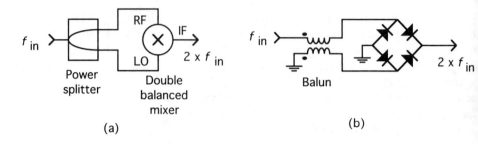

Figure 2.75 Passive frequency doublers.

Active frequency multipliers using bipolar transistors can be designed using the six broad design criteria illustrated in Figure 2.76:

1. Bias the device to be just barely ON in the absence of an input signal.
2. Match the input for maximum power transfer into the device at the fundamental frequency.
3. Provide fully reflective, low-impedance input termination at all harmonic except fundamental.
4. Provide a fully reflective, low-impedance output termination at all harmonics except the desired one.
5. The output circuit should provide a high impedance to the desired harmonic
6. Supply the circuit with sufficiently high input signal level to initiate nonlinear operation.

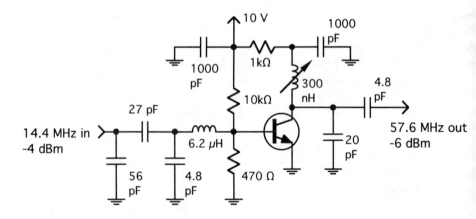

Figure 2.76 Active frequency quadrupler.

The basic idea is to reflect all undesired harmonics back into the device for further conversion into the desired output harmonic. Note that this procedure is the exact opposite to device operation for least distortion, where all undesired harmonics kept away from the active device by nonreflective diplexer terminations. In the example in Figure 2.76, the base circuit is series resonant at 14.4 MHz, and the collector circuit is parallel resonant at 57.6 MHz. Component values can be selected for maximum gain and lower spurious rejection or for lower gain and higher spurious rejection. Spurious here means mainly the ×3 and ×5 products. Using the component values shown in Figure 2.76, all undesired harmonics are more than 18 dB down from the 57.6-MHz signal. Since both input and output circuits are resonant at their respective frequencies, all other harmonics are reflected back into the device. Series-resonant traps called *idlers* are sometimes used in the output circuit to deliberately trap the undesired frequency components; such topology further restricts the operating bandwidth.

Step recovery diode frequency multipliers are used at microwave frequencies to generate high harmonics at a higher efficiency than available from bipolar transistors. Varactor diodes have also been used for frequency multiplication; their relatively high operating impedance makes them suitable for high-power operation. Reference [24] contains a short note on using 180° power splitters and matched FETs to implement a broadband active frequency multiplier with gain, which has inherent suppression of odd harmonics.

2.12 IMPEDANCE-MATCHING NETWORKS

Many system concepts discussed in Chapter 1 require that all available input power be transferred from one stage to the next. Transducer-gain and noise-figure calculations in particular require that all available power be transferred from stage input to its output. This requires that the stages be impedance matched to each other:

The output impedance of a stage and the input impedance of the following stage must be complex conjugates of each other. Interstage impedance match guarantees minimum insertion loss through a chain of devices.

Most impedance-matching problems can be analyzed as trajectories on the Smith Chart, where the addition of a series or shunt component moves the total impedance along constant impedance or admittance circles, as shown in Figure 2.77.

For an example of usage, assume a normalized impedance of $0.3 + j0.5$, which is in the top left quadrant of the Smith Chart. Adding series inductance will move the impedance point clockwise on the impedance chart; adding series capacitance will move the impedance counterclockwise on the impedance chart. Adding parallel inductance will move it counterclockwise on the admittance chart, and adding parallel capacitance will move it clockwise on the admittance chart. The relative amount of movement is given by the equations in Figure 2.77.

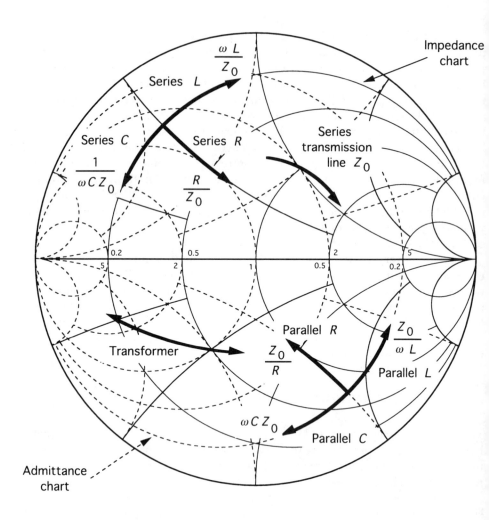

Figure 2.77 Impedance matching guide.

If we want to match our impedance to the reference impedance (usually 50 Ω), then the aim of impedance matching is to arrive at the center of the Smith Chart by traveling along impedance and admittance arcs from the starting point.

If the aim is to provide impedance matching to an impedance other than the reference impedance, then the end point of the matching trajectory is the conjugate of the target impedance. The matching trajectory is plotted on the Smith Chart as each matching component is added.

Constant Q lines, which are defined in Section 4.18, can be overlaid on the Smith Chart shown in Figure 2.78, to estimate the matching network bandwidth. In general, the closer an impedance-matching trajectory comes to the edge of the

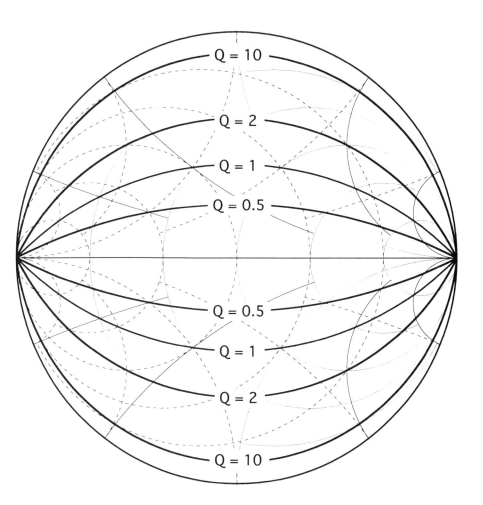

Figure 2.78 Constant Q arcs.

Smith Chart, the narrower the bandwidth. Maximum bandwidth for a given matching network can be obtained by keeping the trajectories short, well away from the edges of the Smith Chart and as close to the real axis as possible. If you want to deliberately design for a certain bandwidth (equal to a certain Q), then make sure that one vertex of the matching trajectory touches the desired constant-Q arc and that all other trajectory points are well inside lower Q regions.

Capacitive tap impedance matching is illustrated in Figure 2.79. A high L/C ratio yields wider bandwidth but introduces greater center frequency error.

$$R_L \approx R_S \left(\frac{C_1 + C_2}{C_1} \right)^2 \qquad (2.48)$$

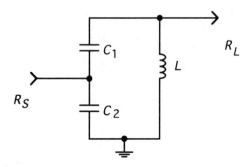

Figure 2.79 Capacitive divider, $R_S < R_L$.

There are two constraints in applying (2.48); the upper limit on L is when C_2 is not required, while at low L/C ratios the circuit becomes sensitive to minute component value changes. As a general rule of thumb, L must be smaller than L_{max} but not very much smaller.

$$L_{max} = \frac{R_L}{\omega\sqrt{\dfrac{R_L}{R_S} - 1}}$$

The tapped coil in Figure 2.18 is also frequently used to achieve impedance transformation; see (2.14).

A single-section quarter-wave transformer can match two resistive impedances. The bandwidth of such a transformer is inversely proportional to the resistance ratio between source and load.

$$Z_0 = \sqrt{R_S R_L}$$

Z_0 = characteristic impedance of $\lambda/4$ line used as impedance transformer

R_S = source resistance (Ω)

R_L = load resistance (Ω)

Multiple $\lambda/4$ sections, shown in Figure 2.80, can be used to widen the bandwidth of the basic, single section $\lambda/4$ transformer.

$$Z_0' = \sqrt[4]{(R_S)^3 R_L}$$

$$Z_0'' = \sqrt[4]{(R_L)^3 R_S}$$

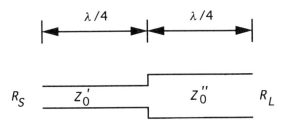

Figure 2.80 Two–section quarter-wave transformer.

The series to parallel transformation in Figure 2.81 is exact at only one frequency and only approximate for wider bandwidths. This approximate narrowband transformation is useful for modeling and matching power amplifier, crystal, and varactor circuits.

$$R_S = \frac{R_P}{1 + \left(\dfrac{R_P}{X_P}\right)^2} \tag{2.49}$$

$$X_S = \frac{\dfrac{R_P^2}{X_P}}{1 + \left(\dfrac{R_P}{X_P}\right)^2} \tag{2.50}$$

$$R_P = R_S\left[1 + \left(\frac{X_S}{R_S}\right)^2\right] \tag{2.51}$$

$$X_P = X_S\frac{1 + \left(\dfrac{X_S}{R_S}\right)^2}{\left(\dfrac{X_S}{R_S}\right)^2} \tag{2.52}$$

Figure 2.81 Series to parallel transformation.

The bandwidth of impedance-matching circuits is governed by the Bode-Fano relationship [25–27], which places an upper limit on such bandwidth. If we want to match a resistive source R to a reactive load $R_L + j\,X_L$, then the bandwidth of the impedance-matching network will be finite, *even if an infinite number of matching components is used!* The bandwidth can be traded off for the desired return loss.

$$\int_0^\infty \ln\left[\frac{1}{\Gamma(\omega)}\right] d\,\omega \le \frac{\pi\,\omega_0}{Q_L} \tag{2.53}$$

Equation (2.53) can be simplified once we realize that the integrated quantity is proportional to the return loss.

$$\text{Return Loss} \le \frac{27.3\,f_0}{Q_L\,BW} \tag{2.54}$$

BW = bandwidth over which return loss (dB) is assumed constant (MHz)

Q_L = load Q, $|X_L/R_L|$

Γ = input reflection coefficient of passive, lossless matching
     network terminated in the load

ω_0 = load resonant frequency; simply a reference frequency at which load
     Q is evaluated and which serves as a reference frequency for
     relative bandwidth expressions.

$f_0 = \omega_0/(2\,\pi)$ (MHz)

There are so many assumptions underlying (2.54) that it is rarely used in practice, but it provides a rough first estimate of achievable bandwidth. The most basic assumption is that the magnitude of the reflection coefficient $|\Gamma|$ is constant over the bandwidth of interest. Another assumption is that the load Q is also constant with frequency.

The limitations imposed by the Bode-Fano relationship can be circumvented by using diplexer or multiplexer networks to break the frequency band into separate bandwidths, over which the Bode-Fano relationship can be individually satisfied.

2.13 MIXERS

Mixers are the primary frequency-translation devices in a communication system. In contrast to frequency multipliers and frequency dividers, which also change signal frequency, mixers faithfully preserve the amplitude and phase properties of signals at the RF port. Signals can therefore be translated in frequency without affecting their modulation properties. AM, FM, and PM modulation indices are preserved in the frequency translation. Important mixer properties are conversion loss, intercept point, LO to RF isolation, and LO noise rejection. Other mixer properties of lesser importance are RF to IF isolation, high-order spurious response rejection, and image noise suppression.

2.13.1 Passive Mixers

The complexity of a mixer depends on its desired performance level. The circuits range from the simplest single-diode mixer to complex image-reject or SSB mixers. Single-diode mixers are often used at high microwave frequencies simply because other alternatives are not available or are too costly.

Port-to-port isolation of single-device mixers depends on external filters; there is no LO AM noise rejection and no inherent suppression of spurious responses. Singly balanced mixers have high LO-RF isolation, RF-IF isolation depends on filters, and even-order LO harmonics are suppressed. Double-balanced mixers have high port-to-port isolation and excellent LO wideband AM noise rejection and suppress even harmonics of both RF and LO signals. Image-rejection mixers provide about 20 dB of image rejection. Figure 2.82 shows the two most widely used passive mixers.

The double-balanced version may or may not contain the RC networks. They are used to improve IP3 in high-performance mixers. Resistor R and local oscillator power are related: higher R requires higher LO power. Capacitor C is an RF bypass, and T_1 and T_2 are 4:1 balun transformers, often implemented as shown in Figure 2.22 at high frequencies. The mixer shown in Figure 2.82(a) is more broadband, because the bandwidth of the single-balanced version of Figure 2.82(b) is limited

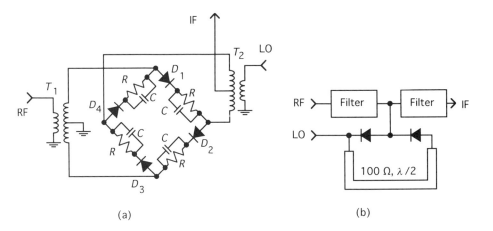

(a)   (b)

Figure 2.82 Schottky diode mixers: (a) double-balanced, (b) single-balanced.

by the $\lambda/2$ length line. For best operation in both versions, all diodes should be matched.

Double-balanced passive mixers have gained popularity because of their high performance and steadily declining cost. The major disadvantage is the high LO power required for linear performance. The input IP3 and the LO power are related by a simple formula (quantities are in decibels): IP3 ≈ (LO power + conversion loss). A given mixer will reach an intercept point plateau at high LO powers, and better IP3 can no longer be obtained by increasing the LO power. In that case, resistor R has to be increased, or higher barrier diodes should be used. The rejection of wideband AM noise on the local oscillator is of particular concern in high-sensitivity receivers and is an important benefit offered by all double-balanced mixers. Section 3.8 deals with this topic in more detail.

The RF termination at the mixer image frequency can be manipulated to improve the conversion loss [28] by a technique known as image enhancement. This practice is discouraged, however, for broadband mixers, because the required image termination cannot be maintained over a wide enough frequency range, and it strongly affects the IF port output impedance. Mismatched reactive terminations at mixer ports are undesirable because they affect IM and spurious rejection in unpredictable ways. A diplexer network on the IF port is mandatory for best performance to avoid reflecting signals back into the mixer for remixing.

The circuit in Figure 2.83 can be used to suppress the image response by about 20 dB, with 10° phase and 1-dB amplitude balance in the circuits.

2.13.2 Active Mixers

Active mixers have the advantage of lower conversion loss (possibly conversion gain) and lower local oscillator power requirement. JFETs are used at lower frequencies, and GaAs FETs are popular at VHF and higher frequencies. While active mixers can have conversion gain, their noise figure is not all that much better than

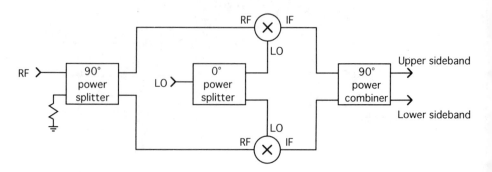

Figure 2.83 Image-reject mixer topology.

diode mixers and hovers around 6 dB. Commercial ICs with integrated low-noise amplifier are available to minimize the impact of poor mixer noise figure, but most integrated amplifier-mixer combinations suffer from poor IM performance.

Single-device mixers have the distinct disadvantage of not attenuating local oscillator AM noise and therefore always require an injection filter. Representative FET circuits are shown in Figure 2.84; bipolar circuits are very similar.

Single-device mixers need to follow some general design guidelines for best performance. In reference to Figure 2.84(a), the source (LO) node should be a short circuit at the RF and IF frequencies for maximum conversion gain, while the gate (RF) node should be a short circuit at the LO frequency to obtain the most transconductance variation due to the LO signal and to prevent LO leakage into the RF port. Conduction of LO energy backward into the receiver's antenna is a concern for most receivers. To prevent oscillations, the drain node should be a short circuit at both LO and RF frequencies, and the gate node should be a low impedance at the IF frequency. This also ensures that noise at the IF frequency is not amplified and added on to the output. The drain node impedance at the IF frequency should be several tens of kΩ for best conversion gain; its value will also affect mixer IM distortion performance. The drain circuit must also have enough rejection of the LO frequency in order not to overload any IF amplifiers further down the chain.

These impedance constraints inherently require that single-device mixers be narrowband devices.

The circuit in Figure 2.84(b) avoids some of the impedance restrictions by providing inherent isolation between the RF and LO signals by the dual gate structure. Both circuits are very sensitive to the drain impedance at the IF frequency. A high impedance, as may result at the edges of a crystal filter, will drastically

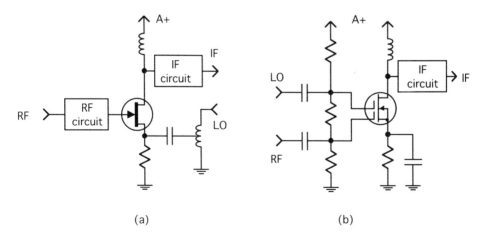

(a)   (b)

Figure 2.84 Single-device active mixers: (a) JFET, (b) dual-gate GaAs.

degrade IM performance. Diplexer or impedance inverter circuit (LC $\lambda/4$ line similar to Figure 2.126) should be used to avoid this difficulty. Double-balanced active mixers have very good performance at low LO levels, but their circuit complexity is higher than passive mixers because three baluns with impedance transformation are required. Another requirement is dc bias. Typical circuits are shown in Figure 2.85 and Figure 2.86.

Transformer baluns T_1 and T_2 are 4:1, while T_3 can be as much as 25:1 for 50-Ω operation due to the very high drain impedance. All capacitors perform RF bypass and dc blocking functions. The four matched FETs can often be obtained in one package, but the disadvantage is the relatively low overall power dissipation, which limits the LO power and therefore the intercept point. With 4-V drain supply, 10-mA bias current, and +7-dBm LO power, the conversion loss is about 1 dB, noise figure is 5.5 dB, and IP3 is near +22 dBm over a broad frequency range.

Component values and design constraints for the circuit in Figure 2.86 are similar to the circuit in Figure 2.85. Resistors R_1 and R_2 provide bias for the top gate (which is the LO input), and resistors R_3 and R_4 provide bias for the other gate. Reference [28] contains a thorough analysis of active and passive mixers of many different topologies. Integrated circuit mixers suitable at frequencies below VHF use an emitter-coupled stage whose current is controlled by the LO signal, as shown in Figure 2.87. This mixer does not reject AM noise on the LO signal because it is only a singly balanced topology.

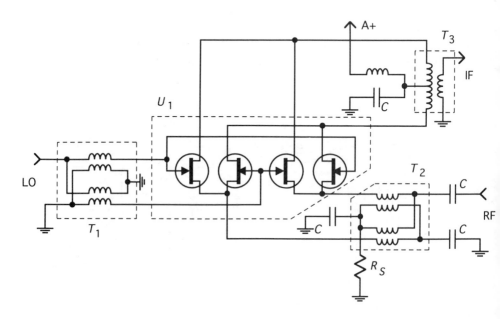

Figure 2.85 Double-balanced active mixer using four matched FETs.

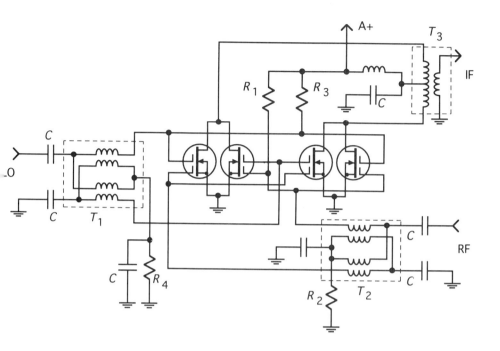

Figure 2.86 Dual–gate FET double-balanced mixer.

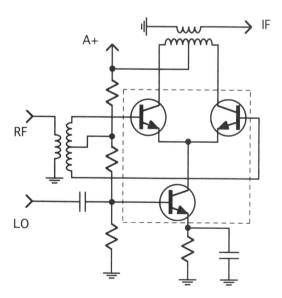

Figure 2.87 Single–balanced mixer using bipolar transistors.

Four-quadrant multipliers based on the Gilbert gain cell are gaining popularity as mixers at lower frequencies [29,30]. Such mixers rely on close matching of their transistors' properties and are therefore only realizable as integrated circuits. Four-quadrant multipliers are linear with respect to both RF and LO ports and therefore do not offer wideband AM noise rejection. Three baluns are required to allow unbalanced operation because the basic multiplier is balanced with respect to ground at all three ports. Emitter-coupled stages used as internal, active baluns typically restrict the operating bandwidth of such devices.

2.14 OSCILLATORS

Oscillators generate the RF signals that are ultimately transmitted by antennas or by a cable system and generate local oscillator signals used in processing receiver signals. The requirements for oscillators used in receivers are more stringent than requirements for transmit oscillators. In addition to frequency stability, receiver local oscillators must have low SSB phase noise required for adjacent channel selectivity and low wideband noise required for good receiver sensitivity and must be free of spurious modulation. Transmitter oscillators do have one unique requirement, called *load pull*, which is a measure of how much the oscillator frequency shifts as a result of changing load conditions such as might be present during transmitter turn-on (see Subsection 1.2.2). In some applications the oscillator's output harmonic content, current consumption, operation over extended temperature range, fast turn-on and turn-off times, and easy modulation capability may be additional requirements.

2.14.1 Crystal Oscillators

Crystal oscillators have two distinguishing characteristics: frequency stability and excellent SSB phase noise. Their drawbacks are low frequency of operation and low-output power level.

Crystal oscillators are basically of two types: fundamental and overtone. The overtone circuits either resonate the crystal parasitic parallel capacitance or provide frequency-selective feedback in order to force the circuit to oscillate at the overtone frequencies. A large variety of crystal oscillator circuits have been investigated [31]; this section presents the most commonly used circuits.

Figure 2.88 and Figure 2.89 show fundamental mode topology, while Figure 2.90 shows a third overtone circuit; all use 16.7-MHz fundamental mode crystal.

Capacitor C_1 can be used to trim the oscillation frequency up in both Colpitts and Pierce designs; an inductor in place of C_1 pulls the frequency lower. In the overtone circuit in Figure 2.90, the resonance formed by L_1 and C_1 must be near the desired operating frequency. The frequency cannot be pulled as easily as in

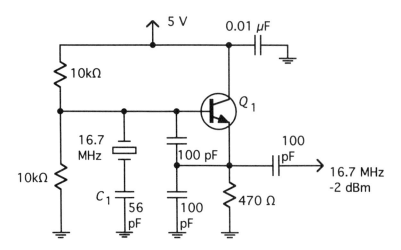

Figure 2.88 Colpitts fundamental mode oscillator.

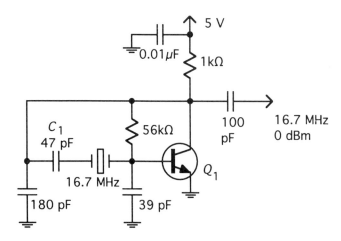

Figure 2.89 Better-performance Pierce fundamental design.

the fundamental mode circuit. A capacitor in series with the crystal will pull the frequency higher, as in the fundamental mode circuit.

Since the overtone frequencies are not exact odd multiples of the fundamental, a significant frequency error may result when a crystal specified at its fundamental mode is used in an overtone circuit. When a crystal is used in the overtone mode, it is best to specify its series-resonant frequency at the desired overtone for best frequency accuracy. Careful measurement of crystal insertion loss in a swept-frequency measurement shown in Figure 2.91(a) will detect the overtone frequen-

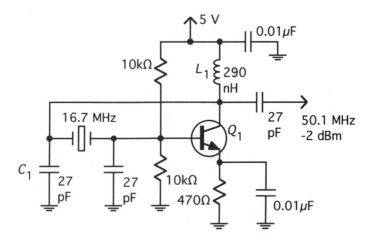

Figure 2.90 Pierce third-overtone oscillator.

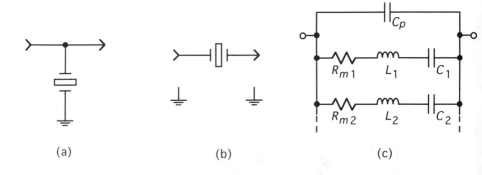

Figure 2.91 Measurements of fundamental and overtone crystal parameters: (a) measuring series resonance, (b) measuring series resistance, and (c) equivalent circuit.

cies from measurement of series resonances and indeed will indicate if a certain overtone can be used at all. If the series resonance is not pronounced (less than a 5-dB notch in a swept-frequency measurement), then an overtone oscillator at that frequency will be difficult to design. Direct measurement of crystal parameters requires a stable, low-noise source, because very narrowband features need to be resolved.

The measurement of insertion loss at series resonance, shown in Figure 2.91(b), can be used to calculate R_m, which is important in estimating crystal power dissipation. Quartz crystals are fragile, and maximum power dissipation specification of less than 100 μW is not uncommon. The crystal power dissipation should be kept low, in order not to affect long-term aging and frequency drift. A few mW of dissipated power will damage the crystal. Therefore, the dc bias of the oscillator transistor should be near 10 mW or less; bipolar devices have significantly higher gain than FETs at such low bias levels and are therefore the preferred active elements. Nevertheless, for very low noise designs, JFETs are still preferred due to their lower flicker noise.

2.14.2 LC Oscillators

Figures 2.92 through 2.94 show basic, representative oscillator circuits; both inductor and capacitor values are inversely proportional to frequency, so that frequency scaling to other frequencies is fairly straightforward, as long as the device has enough gain to ensure oscillation. At high frequencies, the device internal reactances also become important; use of surface-mount chip components is highly recommended. It is also important to ensure that RF chokes and bypass capacitors are suitable for the intended frequency of operation. For example, a 0.01-μF capacitor is not an appropriate bypass at 400 MHz because of its inevitable series inductance. Figures 2.92 and 2.93 show simple Colpitts and Clapp oscillators built with commonly available parts. The virtue of the circuit in Figure 2.93 is that it uses only two capacitance values and a length of standard 50-Ω coaxial cable as a high-Q resonator. Note that the 290-nH inductor is part of the resonant circuit in the 35-MHz oscillator but is just an RF choke in the 700-MHz circuit. The ratios of component values are not very critical, since the circuits will oscillate over a wide range of component values; any of the low-value capacitors or inductors can be adjusted to tune the oscillator frequency.

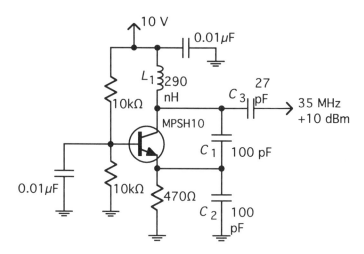

Figure 2.92 Colpitts 35-MHz oscillator.

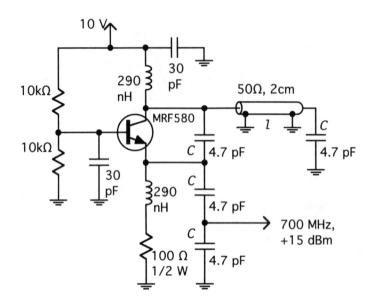

Figure 2.93 Clapp 700-MHz oscillator.

A field effect transistor (FET) oscillator circuit is shown in the subsection 2.14.3; the varactors can be replaced by a capacitor for fixed-frequency operation. In both crystal and *LC* circuits, high *C* to *L* ratios are desirable, so that the external capacitors are much larger than the active device parasitic capacitances, ensuring that the sensitivity of circuit performance to active device variations is minimized. The Inversion and Franklin topologies [32] are also suitable for highly stable *LC* oscillator circuits.

A particularly interesting high-frequency oscillator design is the balanced oscillator shown in Figure 2.94. The base-to-emitter capacitors cause each device to exhibit negative resistance, as well as some reactance. The oscillation frequency is the frequency at which the device input reactances and the transmission line provide a 180° phase shift. The two balanced and out-of-phase outputs can drive a balanced mixer directly, without requiring a balun. When the two outputs are combined using a 180° power combiner, a theoretical 3-dB improvement in SSB phase noise (over that available from a single device) results. The RF voltages are correlated and therefore combine to produce a 6-dB-higher output signal, whereas the SSB phase noise from the two devices is not correlated and produces only 3-dB-higher noise when combined. The result is a theoretical 3-dB improvement in *S/N* ratio compared to a single-device oscillator. If the two outputs are combined using a 0° power combiner, the fundamental frequency signal cancels, while the second harmonic output is enhanced, allowing the generation of higher frequencies using the same basic topology.

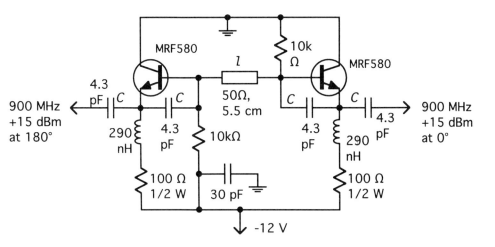

Figure 2.94 Balanced 900-MHz oscillator.

The circuit in Figure 2.94 is very suitable for high-power oscillators: 1-W combined output at frequencies well into the GHz range is feasible with MRF580 devices. The midpoint of the 50-Ω line is a virtual ground and can be used to bring in varactor bias, if required. Since the circuit is balanced, it has inherent suppression of odd harmonics, as long as care is taken to keep the two parts of the circuit electrically symmetrical. A series *LC* circuit can be used instead of the transmission line at lower frequencies, but the simple base to emitter capacitor will fail to produce negative resistance required for oscillation at low frequencies.

Designing an oscillator for particular output power level is not an easy task. There is general disagreement in the literature on how much RF power can be obtained from the supplied dc bias; top estimates vary from 10% to 55% of dc input power, with the highest efficiency available only for low-power circuits (author's own research into solar-powered transmitters). Excess loop gain can result in higher RF power but at the expense of harmonic content and SSB phase noise. Active bias can also help reduce dc power dissipation in bias resistors. Without any special precautions taken, 10% efficiency can be expected. If stable output power level over temperature extremes is desired, then some form of AGC should be used to ensure oscillator startup at temperature extremes.

2.14.3 Voltage-Controlled Oscillators

Voltage-controlled oscillators (VCOs) are most often used as important components of frequency synthesizers and phase-locked loops, less often as swept-frequency signal sources for radar, and for communication systems where the transmitted signal needs to be swept in frequency because the receiver's frequency is unknown, or is unstable.

A VCO design is usually a compromise among a host of conflicting requirements. The primary conflict is between tuning range and SSB phase noise. Some important VCO parameters are operating frequency, tuning range, SSB phase noise, mode hopping (tendency to oscillate at an undesired frequency, sometimes related to L_4), modulation sensitivity, modulation linearity, temperature performance, supply sensitivity, post-tuning drift, spurious output, output power flatness and performance when subjected to mechanical vibration.

Any of the circuits from the previous section can be used as VCOs by replacing a frequency-determining capacitor with a varactor. From the large number of VCO designs, the modified Clapp oscillator shown in Figure 2.95 was selected as a very successful low SSB phase noise design. JFET transistors have the lowest $1/f$ noise but are limited to less than about 600 MHz. Bipolar transistors must be used at higher frequencies.

$C_2 \approx 1/(500\ f)\ \text{(F)}$

$C_3 \approx 1/(500\ f)\ \text{(F)}$

f = oscillator frequency (Hz)

Components C_5, L_1, C_1, and C_4 are selected to produce resonance at the desired frequency; L_1 (possibly including some of C_1) is a high-Q resonator. A high

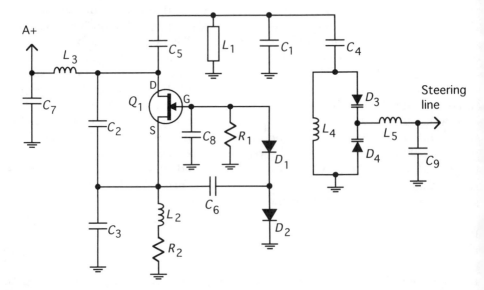

Figure 2.95 Low-noise VCO.

L_1/C_1 ratio is required for wide tuning range. Inductor L_4 across tuning varactors provides compensation if constant steering line sensitivity is required; it also expands the frequency tuning range. Capacitor C_9 can improve SSB-phase noise under the conditions described in Section 4.14. L_3, L_2, and L_5 are RF chokes; C_7, C_8, and C_9 are RF bypass capacitors. Resistor R_2 is selected to be no lower than required to ensure oscillator startup at high temperature. Q_1 is N-channel depletion type JFET, such as the J310. Schottky diodes D_1 and D_2 together with a small value of C_6 provide AGC. The output can be taken from the source node through a small capacitor or at resonator L_1 tap point. Tapping from the resonator has the added advantages of filtering wideband noise and isolating the active device from possible reflection of harmonics from the external environment. Harmonic remixing can degrade SSB-phase noise.

The debate whether to include AGC amplitude control in a VCO has many adherents on both sides. Bipolar oscillators typically have less need for AGC than FET oscillators. You should consider AGC if any of the following operating conditions are encountered:

1. There are fast frequency changes. If the VCO is required to switch over a wide frequency range, AGC may be advantageous. Without AGC, the oscillator may stop oscillating momentarily during the switching interval, and the synthesizer will lose synchronism, with a potentially long recovery interval.

2. Low SSB phase noise designs that operate over a wide temperature range will benefit from AGC, because the gain required to start oscillations at temperature extremes may be too high to ensure low SSB phase noise at room temperature.

3. Constant oscillation amplitude is required. This is especially important in low-noise or wide tuning range designs where it is important to avoid varactor rectification of RF energy. If RF energy is rectified, it changes the dc bias on the varactor, shifts frequency, and increases losses.

4. Oscillator device exhibits large parameter variations from lot to lot.

The AGC widely used with FET oscillators is gate leak bias, where a portion of the RF energy is rectified and the resultant voltage is applied to the gate as a negative voltage to limit device current in inverse proportion to RF amplitude. A simple rectifier for use with a common-gate oscillator is shown in Figure 2.96.

The voltage at the FET gate actually goes below ground as diode D_1 rectifies RF signal present at the source of the device. Resistor R_1 must be large in order not to draw too much current from the rectified RF. Capacitor C_8 is an RF bypass. A voltage multiplier with two diodes and capacitors is sometimes used to increase the AGC range, as shown in Figure 2.95.

If a VCO is designed with internal device capacitances for the feedback or resonator components, then the frequency can be tuned over a wide range simply

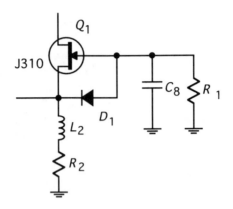

Figure 2.96 AGC scheme used with N–channel depletion–type FET.

by changing the bias current or voltage of the device, without using varactors. In most cases, modulation by power supply variations is generally undesirable, in which case the VCO must have its own separate, well-filtered voltage regulator.

2.15 PLL FILTERS

The loop filter design is governed by three related considerations: lock time, Hum & Noise, and reference-frequency suppression. High loop filter corner frequency typically gives faster lock time, but limits the frequency modulation capabilities of the VCO and yields poor Hum & Noise performance. The loop corner frequency also determines the frequency at which either the VCO phase noise or the reference phase noise becomes important, as mentioned in Section 3.5. In synthesizers with very fine frequency resolution, the reference frequency can be quite low, and the loop filter needs substantial attenuation at this frequency in order not to modulate the VCO with the reference frequency and its harmonics.

Common loop filter topologies are shown in Figure 2.97 through 2.99, together with damping ratio and natural frequency, which are applicable only to second-order transfer function approximations.

The closed-loop bandwidth of the PLL is not the same as the corner frequency of the low-pass filter. The closed-loop bandwidth is made up of all the loop component's contributions. The loop transfer function, $H(s)$, determines the closed-loop bandwidth, ω_n, and damping factor, ζ.

$$H(s) = \frac{\phi_{\text{out}}}{\phi_{\text{in}}} = \frac{K_\phi K_v F(s)}{s + \dfrac{K_\phi K_v F(s)}{N}} \tag{2.55}$$

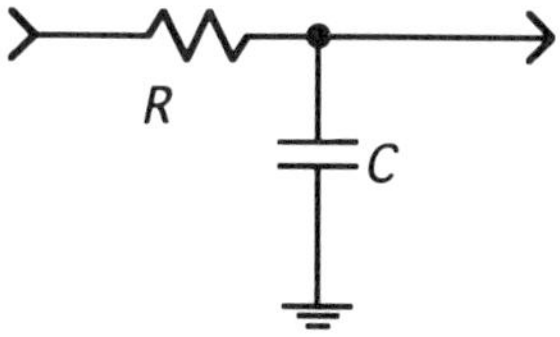

$$F(s) = \frac{1}{1 + s\,R\,C}$$

$$\omega_n = \sqrt{\frac{K_\phi K_V}{N\,R\,C}}$$

$$\zeta = \frac{N\,\omega_n}{2\,K_\phi K_V}$$

Figure 2.97 Passive RC low-pass loop filter.

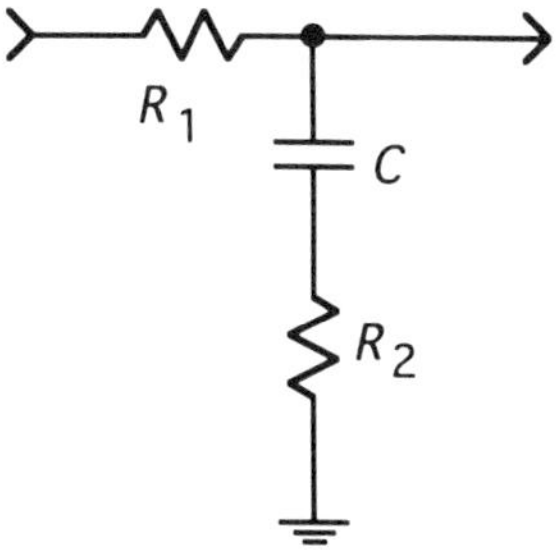

$$F(s) = \frac{1 + s\,R_2\,C}{1 + s\,C\,(R_1 + R_2)}$$

$$\omega_n = \sqrt{\frac{K_\phi K_V}{N\,(R_1 + R_2)\,C}}$$

$$\zeta = \frac{\omega_n}{2}\left(R_2 C + \frac{N}{K_\phi K_V}\right)$$

Figure 2.98 Lead–lag loop filter.

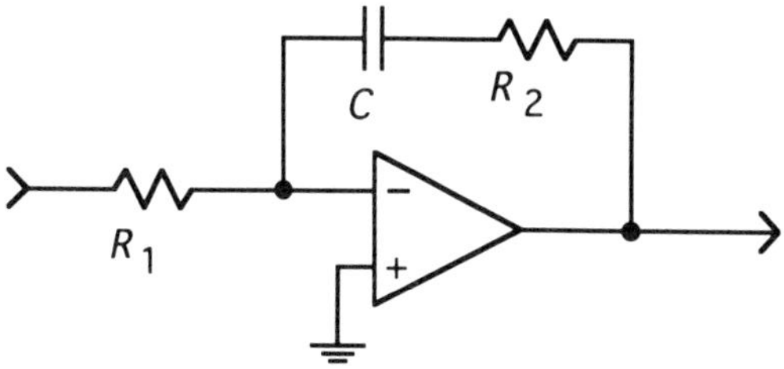

$$F(s) = \frac{1 + s\,R_2\,C}{s\,R_1\,C}$$

$$\omega_n = \sqrt{\frac{K_\phi K_V}{N\,R_1\,C}}$$

$$\zeta = \frac{\omega_n R_2 C}{2}$$

Figure 2.99 Active lead–lag loop filter.

$H(s)$ = closed-loop transfer function

$F(s)$ = loop filter transfer function

ϕ_{out} = phase at VCO output (rad)

ϕ_{in} = phase at phase detector reference input (rad)

K_ϕ = phase-detector conversion factor (V/rad)

K_v = VCO conversion factor, steering-line sensitivity (rad/s/V)

N = frequency divider division ratio; may be noninteger

ω_n = closed-loop natural frequency (rad/s)

ζ = damping ratio of closed-loop frequency response

2.16 POWER SPLITTERS AND COMBINERS

Power splitters are used whenever power needs to be divided or combined while maintaining a good impedance match at all ports. Important power splitter properties are insertion loss, amplitude, and phase balance among the outputs, isolation among the output ports, impedance match at all ports and operating bandwidth. Whether a device operates as a power splitter or a power combiner is determined only by how it is connected; it is the same device.

2.16.1 Resistive Power Splitters and Combiners

The simplest two-way resistive matched power splitter is shown in Figure 2.100.

The resistor values are 16.67 Ω for a 50-Ω splitter, while a 75-Ω splitter requires 25-Ω resistors. The advantage of this circuit is its broad operating bandwidth; disadvantages are the low value (6 dB) of isolation between the output ports and higher insertion loss than can be obtained from reactive power splitters.

2.16.2 Reactive Power Splitters and Combiners

Reactive power splitters have two advantages over their resistive counterparts: lower insertion loss and greater isolation between the ports. The disadvantage is much narrower bandwidth of operation. Figure 2.101 shows the classical Wilkinson power splitter topology.

The 70.7-Ω lines are $\lambda/4$ long in a classical Wilkinson 50-Ω splitter. If we use isolation as a measure of bandwidth, the Wilkinson power splitter can achieve 20 dB or greater isolation bandwidth over a 40% relative bandwidth. Assuming system impedance Z_0, the *TRL* characteristic impedance is $\sqrt{2}\, Z_0$, and isolation resistor R value is $2\, Z_0$.

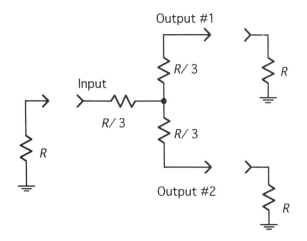

Figure 2.100 A 6-dB resistive power splitter.

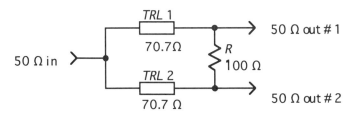

Figure 2.101 Classical Wilkinson power splitter.

A much wider bandwidth can be obtained without increasing the overall length by using the configuration in Figure 2.102 [33]. This fairly elaborate splitter doubles the Wilkinson bandwidth without any physical size penalty. Lengths are in centimeters, resistances in ohms, and capacitances in picofarads. It operates from 460 MHz to 1030 MHz when implemented on a G-10 circuit board. Splitters of this type use multisection nonsynchronous impedance transformers [34] and are best designed by computer optimization techniques.

The 70.7-Ω Z_0 transmission lines in the classical Wilkinson splitter can be approximated by *C-L-C* pi-circuit equivalent networks to arrive at the discrete equivalent circuit shown in Figure 2.103. The bandwidth of this discrete circuit is considerably narrower (by about half) than its transmission line parent:

$$L = \frac{R}{\sqrt{2}\,\pi f}\,[\text{H}]$$

$$C = \frac{1}{2\sqrt{2}\,\pi f R}\,[\text{F}]$$

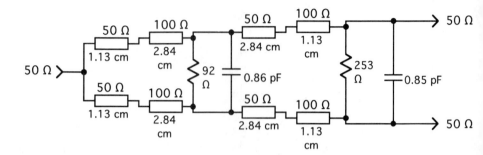

Figure 2.102 Author's broadband splitter, total length $\lambda/4$.

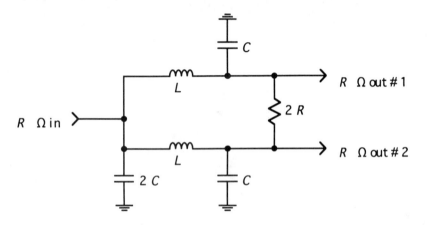

Figure 2.103 Generalized *LC*-equivalent Wilkinson power splitter.

R = system impedance at all three ports (Ω)

f = center frequency of operation (Hz)

A common type of reactive power splitter is shown in Figure 2.104.

L_2 performs the power split, but unfortunately the impedance at point A is $R/2$, so that L_1 is required to perform the impedance transformation from $R/2$ up to the system terminating impedance R. Both L_1 and L_2 should be implemented as transmission-line-type transformers by winding multifilar wire on ferrite toroids. The typical frequency range that can be achieved by splitters of this type is from a few MHz to 1.5 GHz. Figure 2.105 and Figure 2.106 illustrate the winding technique for L_1 and L_2.

For L_2, shown in Figure 2.105, two or three turns of twisted bifilar wire are wound through a suitable toroid, connecting the red and green wires in a series aiding configuration. The junction of red and green wires is the center tap.

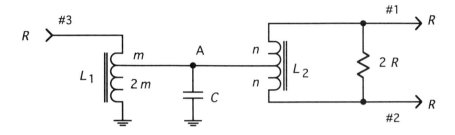

Figure 2.104 Transformer–type power splitter.

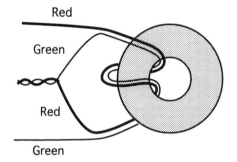

Figure 2.105 L_2 construction.

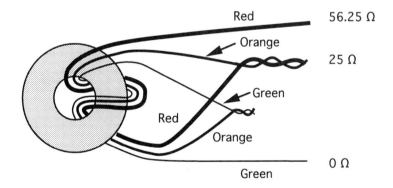

Figure 2.106 L_1 construction.

For L_1, shown in Figure 2.106, two turns of trifilar wire are wound on a suitable toroid, connecting all wires in a series-aiding configuration. This winding scheme transforms 25 Ω to 56.25 Ω rather than 50 Ω, which represents a theoretical VSWR of 1.125 or a return loss of 24.5 dB, which is adequate for most applications.

The value of capacitor C is best determined by experiment; it is usually around 10 pF for a broadband power splitter. The ferrite loss at high frequencies effectively adds shunt resistance to the isolation resistor, so in practice better performance is obtained with a 10% to 20% larger value resistor (120 Ω for a 50-Ω splitter), depending on ferrite characteristics. Better than 20-dB isolation and less than 4-dB insertion loss are easily achieved from 5 MHz to above 500 MHz with this type of construction.

Figure 2.107 and Figure 2.108 show modifications of the basic VSWR bridge topology for use as 6-dB power combiners with high isolation.

For both combiners, $R_1 = R_2 = R_3 =$ system impedance. The isolation between input ports depends only on the quality of the transformer and control of stray capacitances. The isolation between ports 1 and 2 is usually better than in the 3-dB transformer type splitter-combiner in Figure 2.104.

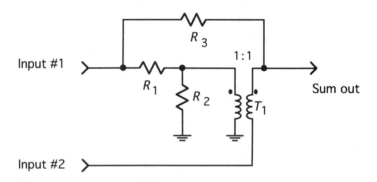

Figure 2.107 A 6-dB power splitter/combiner.

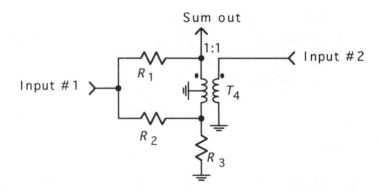

Figure 2.108 VSWR bridge used as a 6-dB power combiner, similar to Figure 2.107.

2.16.3 *n*-Way Power Splitters and Combiners

All n transmission line sections in Figure 2.109 are $\lambda/4$ long and can be realized using any of the lumped-component approximations shown in Section 2.19. The Wilkinson power splitter in Figure 2.101 is a special case with $n = 2$.

The n isolation resistors are equal to the terminating impedances. Their junction is floating but will invariably have some capacitance to ground. The effect of this capacitance is not negligible, because it shifts the isolation and return loss frequency responses in opposite directions.

2.16.4 Unequal Power Splitters and Combiners

The power split between ports 2 and 3 in Figure 2.110 is not arbitrary but is governed by (2.56) and (2.57).

$$IL2 = -20 \log\left(\frac{R}{R_2 + R}\right) \tag{2.56}$$

$$IL3 = -20 \log\left(\frac{R}{R_1 + R}\right) \tag{2.57}$$

$$R = \sqrt{R_1 R_2}$$

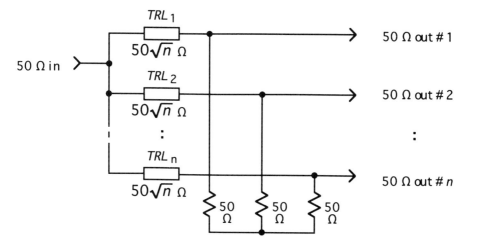

Figure 2.109 Narrowband *n*–way power splitter.

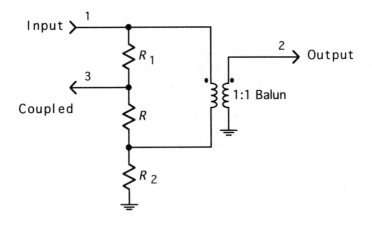

Figure 2.110 Resistive unequal power splitter with isolation.

IL2 = insertion loss from port 1 to port 2 (dB), IL2 > 0

IL3 = insertion loss from port 1 to port 3 (dB), IL3 > 0

R = terminating impedance to ground at all ports (Ω)

See Section 2.9 for derivation of this circuit from a directional coupler.

Narrowband, in-phase, unequal power splitters of the type shown in Figure 2.111 have been investigated [17] and can be realized as transmission line circuits or their lumped-component equivalents, using the transformations of Section 2.19.

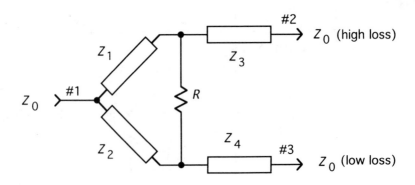

Figure 2.111 Narrowband unequal power splitter topology.

$$k = \sqrt{\frac{P_3}{P_2}} = 10^{(P_2-P_3)/20}, \; k > 1$$

$$Z_1 = Z_0\sqrt{k(1 + k^2)}$$

$$Z_2 = Z_0\sqrt{\frac{(1 + k^2)}{k^3}}$$

$$Z_3 = Z_0\sqrt{k}$$

$$Z_4 = \frac{Z_0}{\sqrt{k}}$$

$$R = Z_0\frac{1 + k^2}{k}$$

k = voltage ratio between ports (linear)

P_2 = relative power at port 2 (W)

P_3 = relative power at port 3 (W)

P_2 = insertion loss to port 2 (dB)

P_3 = insertion loss to port 3 (dB)

Z_0 = system characteristic impedance (Ω)

All line lengths are $\lambda/4$.

The bandwidth and VSWR performance of this unequal power splitter depends on the desired power ratio but is typically near 15%, limited by return loss at port 1. In some cases, the circuit can be simplified by letting one of Z_3 or Z_4 equal Z_0 and reoptimizing the circuit for the desired performance. This has been done in the circuit in Figure 2.112, which shows a lumped-element realization.

This power splitter operates at 165 MHz with 20% relative bandwidth; output 2 is −0.6 dB and output 3 is at −11 dB with 40 dB of isolation between ports 3 and 2. Capacitor C should theoretically be a small value, but it can often be left out because the required capacitance may be supplied by the parasitic capacitance of the connector or circuit board pad at port 2.

Any of the directional coupler circuits discussed in Section 2.9 can also be used as unequal power splitters.

2.16.5 180° Power Splitters and Combiners

The in-phase power splitter covered in Subsection 2.16.2 and the 180° splitter are special cases of the same device, a simple center-tapped transformer sometimes called magic-T, illustrated in Figure 2.113.

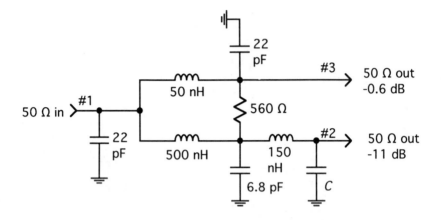

Figure 2.112 Lumped-element unequal power splitter with isolation.

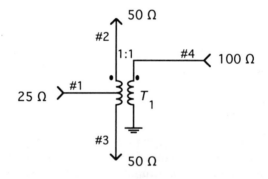

Figure 2.113 Universal power splitter.

In the 0° power splitter, port 1 is the input, and its 25-Ω input impedance is transformed up to 50 Ω by a separate transformer, while port 4 is not used and is terminated internally in a balanced 100-Ω resistor, which therefore does not require the transformer secondary winding. Ports 2 and 3 are the in-phase outputs. To convert this circuit into an 180° power splitter, use port 4 as input, then port 2 will be 0° output and port 3 will be 180° output, both at −3 dB; port 1 will be isolated. The only remaining challenge is to implement the 2:1 impedance transformation in T_1, required at port 4. Or we can leave T_1 as a 1:1 balun and implement the impedance step up by a separate autotransformer at port 4. Transformer T_1 in Figure 2.114(a) can be implemented as 6:4 turns ratio bifilar transformer. The configuration in Figure 2.114(b) may be easier to implement because the transformer construction is much simpler.

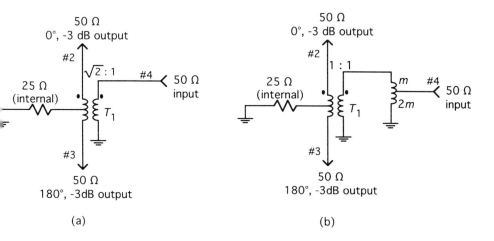

Figure 2.114 Two 180° power splitters using a center–tapped transformer: (a) one-transformer circuit, (b) two-transformer circuit.

The Wilkinson power splitter in Figure 2.101 can be analyzed in exactly the same way: The 100-Ω isolation resistor can be replaced by a balun and will become port 4, the input for a 180° power splitter. If the balun is implemented by the circuit in Figure 2.24(b), the Wilkinson power splitter becomes a circuit known as the *rat–race hybrid*, shown in Figure 2.115. This power splitter can potentially achieve better isolation than the conventional Wilkinson splitter, because the two output

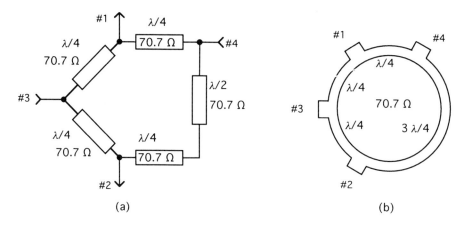

Figure 2.115 Development of rat–race hybrid from Wilkinson power splitter. (a) Wilkinson with transmission line balun, (b) conventional form.

ports are not in close physical proximity, as they are in the Wilkinson splitter, due to the presence of the isolation resistor between them.

All ports of the rat-race hybrid must be terminated in the system characteristic impedance; the termination impedance R at all ports of the rat-race hybrid is the same, and all transmission line characteristic impedances are $\sqrt{2} \cdot R$, or 70.7 Ω in a 50-Ω system. The relative bandwidth of this power splitter is about 20% when used as an in-phase or 180° power splitter. Table 2.1 shows the two modes of operation.

Table 2.1
Rat-Race Hybrid Port Properties

Mode	Input Port	Output Port	Output Port
In-phase splitter	3	1 at 0°, −3 dB	2 at 0°, −3 dB
180° splitter	4	1 at −90°, −3 dB	2 at −270°, −3 dB

2.16.6 90° Power Splitters and Combiners

The 90° power splitter cannot be derived from the magic-T and has unique properties. A transmission line implementation is shown in Figure 2.116.

The bandwidth of this transmission line splitter is limited by isolation between the two output ports. A 3-dB insertion loss, return loss, and 90° phase difference cover a wider bandwidth than isolation. All line lengths are $\lambda/4$. A lumped-element version can be constructed by using pi-circuit C-L-C sections for the $\lambda/4$ transmission lines (see Section 2.19).

A transformer version of the 90° power splitter, similar to a directional coupler, is shown in Figure 2.117.

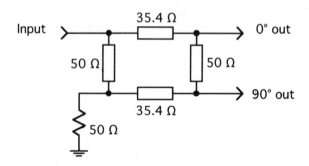

Figure 2.116 Narrowband (10%) 50-Ω 90° power splitter.

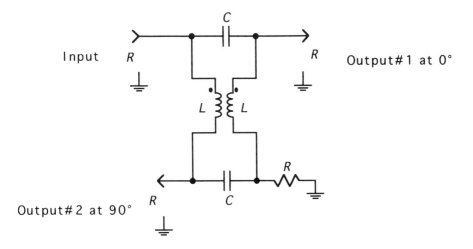

Figure 2.117 Lumped-element quadrature coupler.

$$L = \frac{R}{2\sqrt{2}\,\pi f}$$

$$C = \frac{1}{2\sqrt{2}\,\pi f R}$$

R = system, terminating impedances (Ω)

f = frequency of 3 dB coupling (Hz)

The transformer should be constructed with twisted bifilar wire wound on a ferrite core to achieve close to unity coupling. The bandwidth is limited by the desired accuracy on the 3-dB power split. It is approximately 20% for ±1 dB amplitude balance. If the coupling is less than unity, the operating frequency increases, the bandwidth widens, but the 90° phase balance deteriorates.

Any 3-dB coupled-line directional coupler, such as the one shown in Figure 2.56, can also be used as a 90° power splitter. Special multisection coupled-line couplers, called Lange couplers, have been developed specifically for use as broadband 90° power splitters.

An interesting property of 90° splitters not shared by the other types is that they can be used to approximate a circulator, as shown in Figure 2.118.

The input impedance of this configuration is 50 Ω, regardless of the impedance of circuits A and B, as long as both circuits A and B are exactly the same and the splitters are ideal. This property also applies to the output, so that this technique can be used to work with reflective devices, without introducing any mismatch at either the input or the output.

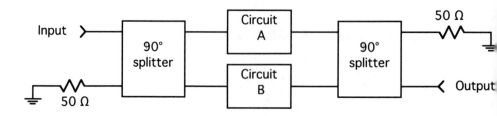

Figure 2.118 Matching of reflective devices.

2.17 SUPERREGENERATIVE RECEIVER

A superregenerative receiver is simply a gated oscillator, an oscillator that is switched on and off at a regular rate. A simplified block diagram is shown in Figure 2.119. The principle of operation relies on the fact that an oscillator normally starts oscillating when thermal noise is amplified in a positive feedback so that the oscillation amplitude builds up from thermal noise and increases until some form of limiting occurs.

When an external signal at the oscillation frequency is applied to an oscillator that has just been switched on, the oscillation amplitude builds up faster, because the oscillator is not starting from thermal noise, but amplifies a signal that is already supplied by the external source. Therefore, the duty cycle of the oscillation amplitude changes in proportion to the amplitude of an externally applied signal, as shown in Figure 2.120.

Many types of superregenerative receivers have been developed over the years. Shaping the rise and fall times of the gating signal can improve the linearity and

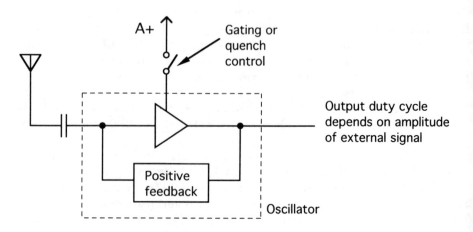

Figure 2.119 Block diagram of a superregenerative receiver.

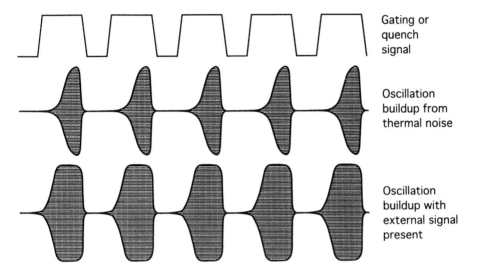

Figure 2.120 Superregenerative receiver waveforms.

dynamic range; when the gating signal is applied externally, the receiver is said to be *externally quenched*. When the biasing is arranged such that the oscillation dies by itself when it reaches a certain amplitude, it is *self-quenched*.

It should be fairly obvious that this type of receiver is suitable only for demodulating amplitude-modulated signals; more specifically, it is very suitable for demodulating OOK data signals. The quench frequency must be much higher than the data rate.

The AM demodulation process can be done externally by an envelope detector and a low-pass filter, but in the absolute lowest cost and simplest designs, monitoring the dc bias of the active device provides the envelope detection shown in Figure 2.121. The bias current changes as the oscillation amplitude builds up. Therefore, a quench method that does not directly affect the bias current needs be devised.

L_2 and C_4 (both large values) provide the self-quenching action. As the oscillation amplitude builds up, the collector current drops, resulting in lower voltage drop across R_4, pulling one end of C_4 toward ground. C_4 is large and it also pulls the base of Q_1 low, decreasing the current even more, which lowers the gain sufficiently to stop oscillation. L_2 is also large and prevents a sudden change of emitter current, so that the current remains low long enough to make sure all energy storage at the oscillation frequency is dissipated before the oscillation can build up from thermal noise again. In this type of detector, the self-quench frequency also changes in response to the input amplitude.

The main purpose of the RF amplifier ahead of the detector is to prevent the oscillation energy from leaking into the antenna. Nevertheless, two receivers

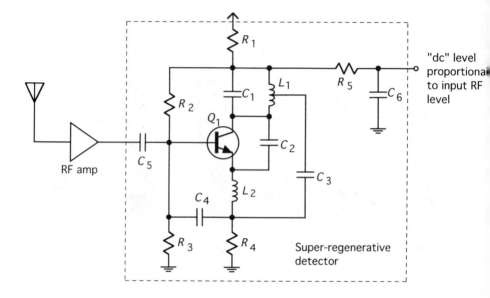

Figure 2.121 Self–quenched superregenerative receiver.

operating in close proximity will interfere with each other because the RF amplifier cannot provide sufficient reverse isolation.

Despite their lowly reputation as remote control garage door openers, superregenerative receivers are actually very clever devices, offering a rich set of design tradeoffs involving current consumption, communication data rate, receiver selectivity, and sensitivity. It is not difficult to construct a receiver with less than 1 mA of current consumption. The receiver bandwidth and selectivity can be controlled by designing the oscillator for a specific loaded Q.

Externally quenched receivers usually have better performance than self-quenched designs.

The main drawbacks of superregenerative receivers are insufficient RF selectivity, susceptibility to overload from strong on- or off-frequency carriers, radiation of oscillator energy, frequency drift with temperature, and interference between receivers in close proximity. Therefore, these receivers are limited to short-range, low data rate applications where the receivers are far apart and frequency drift is not important.

2.18 SWITCHES

The requirements for the switching of RF signals are so diverse that no one circuit can be advocated as a universal RF switch. The following questions can serve as a guide to clarify the switch requirements:

- What is the power level of the switched signal or signals? Are harmonics or IM distortion generated by the switch a concern? RF relays have the best power handling and IM performance.
- Do the off ports need to be resistively terminated? This determines the switch topology. Pay attention to the power-handling requirement of the resistive termination.
- Is the switch likely going to operate with highly reactive loads? This may degrade switch isolation.
- What is the required switching speed? Is switch bounce acceptable? Are there constraints on rise, fall, and settling times? RF relays are much slower than other types of switches.
- What is the desired bandwidth and frequency of operation? This may affect the choice of capacitors and chokes used. PIN diode and GaAs transistor switch IM distortion increases at low frequencies. What is the desired reliability? RF switching relays may not be desirable in high-reliability applications.
- What are the electrical requirements for insertion loss and isolation? Is power consumption a concern?

As a general rule, RF relays are suitable for high-power, low-distortion, low-switching-speed, and wide-bandwidth applications. RF relays usually have very respectable performance. They are rather costly and bulky.

PIN diode switches are generally used for low-power, high-speed, and high-reliability applications. The equivalent circuit of a PIN diode is a small resistor when forward dc biased and a small capacitor when reverse biased.

Gallium arsenide microwave monolithic integrated circuits (MMIC) switches offer a wide range of performance but at a cost premium. Compared to PIN diode switches, GaAs switches can be used at higher frequencies and offer faster switching speed and lower power consumption.

Frequently a simple, properly biased series PIN diode switch performs adequately, but if it is required to switch a highly reactive load, such as a filter, there will be a frequency at which the load inductance will resonate with the diode OFF capacitance, destroying switch OFF isolation. The series-shunt switch shown in Figure 2.122 is suitable for such reactive load applications and should be preferred to a simple series switch whenever switching between filters.

The switching is arranged such that when D_1 is on, D_4 and D_3 are off, and D_2 is on. The RF signal thus passes from the input to output 1, and RF output 2 is connected to ground via D_2. This prevents an inductive impedance at output 2 from resonating with the OFF capacitance of D_3.

A 50-Ω resistor can be placed in series with C_4 if it is required to terminate the OFF port in 50 Ω.

The basic transmit-receive switch in Figure 2.123 shows that D_1 and D_2 are either both ON (transmit) or both OFF (receive). Receiver isolation when transmitting is achieved by using D_2 as a short-circuit termination for the $\lambda/4$ section, which then

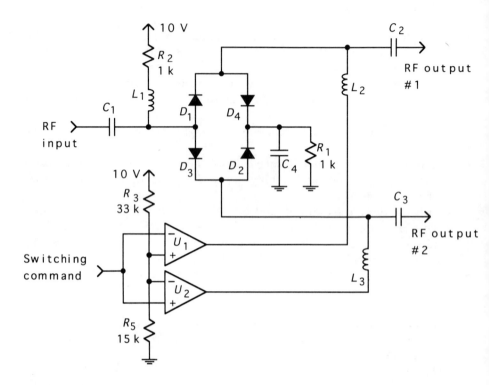

Figure 2.122 PIN diode series–shunt switch.

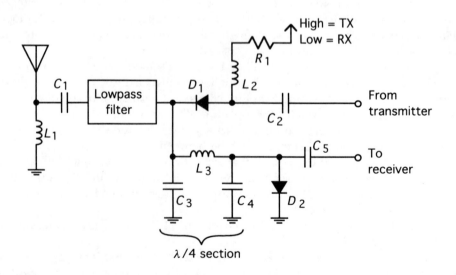

Figure 2.123 Transmit-receive switch for a simplex transceiver.

appears as an open circuit when one looks toward C_3 from the antenna. The design of $\lambda/4$ section by an LC pi-circuit is illustrated in Section 2.19.

C_1, C_2, and C_5 are dc blocking capacitors, while C_3 and C_4 have values that depend on the operating frequency. Similarly, L_1 and L_2 are RF chokes, while the value of L_3 also depends on the operating frequency. The purpose of L_1 is to provide a dc path to ground for the antenna, to prevent static voltage buildup from destroying C_1 and damaging the PIN diodes.

Another $\lambda/4$ section terminated by a PIN diode can be used on the receive side to improve the transmit-to-receive isolation in high-power transmitters, as shown in Figure 2.124. Resistors R_2 and R_3 ensure that both D_2 and D_3 have the same current; C_8 and C_9 are RF bypass capacitors. Capacitors C_4 and C_6 can be combined into one capacitor, equal to their sum.

It may appear that C_4 and C_7 are shunted by the PIN diodes to ground and are thus not required. Consider the case when the diodes are off in receive mode: without C_4 and C_7, the $\lambda/4$ sections would not really be $\lambda/4$ sections, and the LC network would introduce a large mismatch, instead of being just an equivalent length of matched transmission line.

Resistor R_1 sets the dc bias current through the PIN diodes, which in turn determines the level of harmonics that the switch will generate in transmit operation. Most applications require a low-pass filter between the switch and the antenna in order to suppress diode-generated harmonics. The amount of harmonics and IM distortion are related to a PIN diode property called transit time. In general, a longer transit time diode is required for low-frequency or low-distortion applications [35].

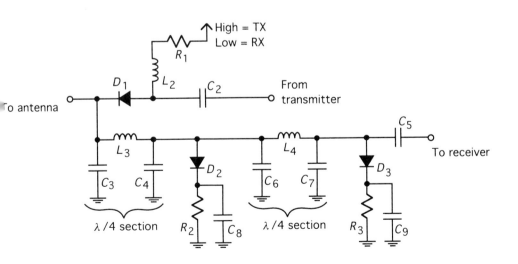

Figure 2.124 Switch with improved transmit-to-receive isolation.

All of these switches require a dc level change to activate the switch. While the RF circuits are isolated from the dc circuits, a step change in voltage across a dc blocking capacitor will introduce a current spike into the RF circuits. This is especially undesirable for fast-acting switches at lower frequencies, where the coupling capacitors need to be large. The circuit in Figure 2.125 does not change any dc levels across RF blocking capacitors and is therefore suitable for very fast switching at lower frequencies.

Single FET and bipolar transistors can also be used to perform switching functions, with an additional design constraint: The circuit must be stable when transistors are between their two switching states. An oscillating switch stage may latch and remain oscillating despite a change in its bias voltage brought about by the switching command.

The bias current requirements for a PIN diode switch are dictated by the desired ON resistance, and the generated IM distortion and harmonic levels. In small-signal applications (RF levels below −10 dBm), the low ON resistance requirement dominates, and a switch with 10-mA bias will have adequate IM performance. In high-power switching applications, the ON resistance is still important, but IM and harmonics will need careful attention. In general, the carrier lifetime of the particular diode will have as much of an effect as the bias current and is responsible for a 6-dB-per-octave *improvement* in IM distortion as the operating frequency is increased. There is a 10-dB improvement in IM products for every doubling of dc bias current between approximately 1 mA and 10 mA. Above 10 mA, the relationship between IM and bias current tends to be linear.

Typically, bias current of 50 mA is required to switch 25W of RF power in a well-designed PIN diode switch. A high-power switch should be followed by a low-pass filter to attenuate any harmonics generated in the PIN diodes. Component value tolerances, temperature variations, power dissipation, and high VSWR in particular have to be taken into account in the design of high-power, low-distortion PIN diode switches.

GaAs switches are monolithic integrated circuits incorporating two or more FET transistors to realize a series-shunt switch. Since the negative switching signal

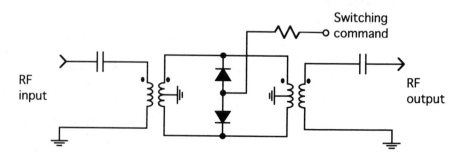

Figure 2.125 Noise–free RF switch using matched diodes.

controls a high-impedance FET gate, and no dc bias current is required, the current consumption is very low. Typical performance of an integrated GaAs switch (MGS-70008) is operating bandwidth dc to 3 GHz, 1-dB insertion loss, 30-dB isolation, maximum RF power of +15 dBm, +45 dBm IP3, and 3-nS switching speed.

2.19 TRANSMISSION LINE SECTION

A $\lambda/4$ length of transmission line can be approximated by any of the circuits in Figure 2.126 or Figure 2.127. Which one of the four circuits to use depends on the actual application:

$$L = \frac{Z_0}{2\,\pi f} \tag{2.58}$$

$$C = \frac{1}{2\,\pi f Z_0} \tag{2.59}$$

Z_0 = characteristic impedance of transmission line (Ω)

f = frequency at which line is $\lambda/4$ long (Hz)

L = inductance (H)

C = capacitance (F)

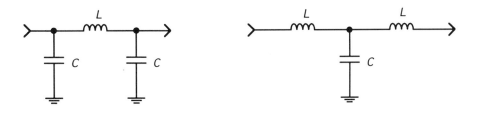

Figure 2.126 Low–pass pi–circuit and T–circuit approximations to a $\lambda/4$ length transmission line.

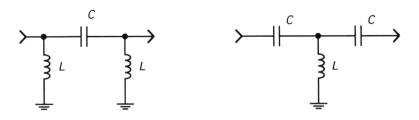

Figure 2.127 High–pass pi–circuit and T–circuit approximations to a $\lambda/4$ length transmission line.

2.20 VSWR BRIDGE

A VSWR (voltage standing wave ratio) bridge is a device that measures incident and reflected voltage for any load connected to the test port. It is basically a Wheatstone bridge circuit, shown in Figure 2.128.

All the resistors are equal to the system impedance, 50 Ω in our case. The voltage across R_i is proportional to the incident voltage across the device under test (DUT), while the voltage across R_r is proportional to the reflected voltage from the DUT. Despite its apparent simplicity, the VSWR bridge is frequently a complicated device in practice, because both these voltages are balanced and need to be transformed by good baluns to unbalanced voltages for external measurement. If phase information is not important, then the baluns can be dispensed with, and detector diodes can be mounted in the appropriate locations to convey dc voltages proportional to RF levels to the measurement environment. Figure 2.129 illustrates the use of 1:1 baluns to bring both incident and reflected signals into an unbalanced measurement environment.

The theoretical insertion loss is 6 dB from input to incident and 9 dB from input to reflected. The reflected port measures return loss directly, when first properly calibrated with an open and a short at the DUT terminals.

The main challenge in constructing this type of bridge is the broadband 1:1 balun performance. The universal power splitter in Figure 2.113 can also be used as an impedance bridge; if port 4 of that figure is used as input and both ports 2 and 3 are properly terminated, then port 1 is isolated. Any impedance imbalance between ports 2 and 3 shows up as measurable signal at port 1. That signal is theoretically equal to the return loss. Therefore, a 180° power splitter can also be

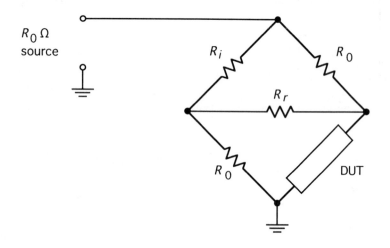

Figure 2.128 VSWR bridge.

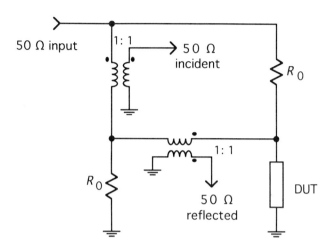

Figure 2.129 VSWR bridge suitable for unbalanced measurement environment.

used for measurement of return loss, by using port 2 as reference and port 3 as the unknown impedance port.

In-phase power splitter in Figure 2.104 will measure return loss if port 1 is input, port 2 is output, and port 3 is connected to the unknown impedance.

Directional couplers and power splitters can be used to measure return loss, provided they are properly calibrated with open and short terminations, and their directivity and isolation respectively are high enough for an accurate measurement. Refer to Section 3.3 for determining the effect of measurement setup directivity on return loss measurement.

REFERENCES

[1] Porter, J. "Multiple-Tuned Amplifiers," *RF Design*, Jan.-Feb. 1983, p. 38.

[2] Porter, J. "Stagger-Tuned Bandpass Active Filters," *RF Design*, Mar. 1988, p. 39.

[3] Tsironis, C., *Load Pull Measurements on Very Low Impedance Transistors*, Focus Microwaves Inc. Application Note No. 6, Pointe-Claire, Quebec, Canada, Nov. 1993.

[4] Motorola Application Note #AN555, *Mounting Stripline–Opposed–Emitter (SOE) Transistors*, RF Device Data, Vol. 2, 1988, pp. 7–65.

[5] MiniCircuits, "Mixer Handbook," sect. 2 of *RF/IF Signal Processing Handbook*, Vol.1, Brooklyn, NY, 1986.

[6] Stubbs, G. "New Application of Old Transformer Theory," *Communications Engineering and Design*, June 1986, p. 16.

[7] Johnson, R. C., and H. Jasik, *Antenna Engineering Handbook*, 2d ed. New York: McGraw-Hill, 1984, pp. 43–23.

[8] American Radio Relay League, *The ARRL Handbook for Radio Amateurs*, Newington, CT, 1992, pp. 2–17.

[9] Cloette, J. H., "Exact Design of the Marchand Balun," *Microwave Journal*, May 1980, p. 99.

[10] Motorola Semiconductor Application Notes AN593, AN721, AN749, AN1034, and EB104.

[11] Kinsman, R. G., *Crystal Filters. Design, Manufacture, and Application*. New York: John Wiley & Sons, 1987, p. 66.

[12] Lathi, B. P., *Communication Systems*. New York: John Wiley & Sons, 1968, p. 189.

[13] Young, P. H., *Electronic Communication Techniques*, 2d ed. Columbus, OH: Merrill Publishing, 1990, pp. 374–382.

[14] Carlson, A. B., *Communication Systems*, 2d ed. New York: McGraw–Hill, 1975, pp. 249, 658.

[15] Young, P. H., *Electronic Communication Techniques*, 2d ed., Columbus, OH: Merrill Publishing, 1990, p. 324.

[16] Methot, F., "Constant Impedance Bandpass and Diplex Filters," *RFDesign*, Nov. 1986, p. 92.

[17] Howe, H., *Stripline Circuit Design*. Dedham, MA: Artech House, 1974, pp. 97, 159.

[18] ingSOFT Limited, *RFLaplace*, 213 Dunview Ave., Willowdale, ONT M2N 4H9, Canada, (416) 730-9611.

[19] Hewlett-Packard, *Momentum*, Santa Rosa Systems Division, 1400 Fountaingrove Parkway, Santa Rosa, CA 95403.

[20] Ansoft Corporation, Four Station Square, Suite 660, Pittsburgh, PA 15219, (412) 261-3200.

[21] Humpherys, D. S., *The Analysis, Design, and Synthesis of Electrical Filters*. Englewood Cliffs, NJ.: Prentice-Hall, 1970, p. 466.

[22] Zverev, A. I., *Handbook of Filter Synthesis*. New York: John Wiley and Sons, 1967, p. 516.

[23] Vizmuller, P., "An Absorptive Notch Filter," *RF Design*, July 1988, p. 31.

[24] TOKO, *Balun Transformers for Mixers, Frequency Multipliers, and Broadband Impedance Matching*, Application Note CF-119, TOKO America, 1250 Feehandville Dr., Mount Prospect, IL 60056.

[25] Bode, H. W., *Network Analysis and Feedback Amplifier Design*. New York: D. van Nostrand, 1945.

[26] Fano, R. M., "Theoretical Limitations of the Broadband Matching of Arbitrary Impedances," *Journal of the Franklin Institute*, Vol. 249, Jan.-Feb. 1950, pp. 57–84, 139–154.

[27] Matthaei, G., L. Young, E. M. T. Jones, *Microwave Filters, Impedance–Matching Networks, and Coupling Structures*. Dedham, MA: Artech House, 1980, p. 3.

[28] Maas, S. A., *Microwave Mixers*. Dedham, MA: Artech House, 1986, p. 120.

[29] Gilbert, B., "A Precise Four-Quadrant Multiplier With Subnanosecond Response," *IEEE J. Solid State Ckts.*, SC-3(4), Dec. 1968, pp. 365–373.

[30] Grebene, A. B., *Analog Integrated Circuit Design*. New York: Van Nostrand Reinhold,1972, ch. 7.

[31] Matthys, R. J., *Crystal Oscillator Circuits*. New York: Wiley Interscience, 1983.

[32] Brown, F., "Stable LC Oscillators." In *Oscillator Design Handbook*, a collection from *RF Design* magazine, p. 76.

[33] Vizmuller, P., "Broadband Miniature Power Splitter," *RF Design*, Aug. 1987, p. 49.

[34] Matthaei, G., L. Young, E. M. T. Jones, *Microwave Filters, Impedance–Matching Networks, and Coupling Structures*. Dedham, MA: Artech House, 1980, p. 335.

[35] *PIN Diode Designers Guide*, Microwave Associates, Inc., Burlington, MA, 1980, p. 34.

CHAPTER 3

Measurement Techniques

3.1 ANTENNA GAIN

Antenna gain is the ratio between maximum radiation intensity and radiation intensity from an isotropic (omnidirectional) antenna supplied with the same power. *Directivity* is defined as the ratio between maximum radiation intensity and average radiation intensity. Gain includes antenna losses and inefficiency; directivity does not. *Polarization* refers to the orientation of the electric vector in the radiated wave; a dipole antenna produces polarization parallel to the long axis of the antenna. It does not radiate and will not receive a wave polarized perpendicular to its long axis.

Antenna gain can be estimated by measuring maximum far-field transmission between two identical antennas a fixed distance apart. The polarizations must match, and there should be no reflections. Adjust the two antennas relative to each other for maximum transmission.

$$G = \frac{4\pi d}{\lambda}\sqrt{\frac{P_r}{P_t}} \tag{3.1}$$

G = antenna gain over isotropic (linear)

d = distance between two antennas, $d > 10\lambda$ (m)

P_r = received power (W)

P_t = transmitted power, power fed into antenna (W)

λ = free-space wavelength (m)

This measurement gives gain, not directivity, because it includes the effect of antenna radiation efficiency.

3.2 COMPONENT VALUE MEASUREMENTS

3.2.1 Capacitors and Inductors

There are many specialized instruments that, when properly calibrated, can measure specific component values. The simple procedures described in this section do not require specialized equipment and yield component values by measuring frequency response. Most component measurement instruments are limited to less than 500 MHz in their capability and are not specifically designed to measure component parasitics. Direct impedance measurements using the Network Analyzer also need to be converted to the corresponding combination of component and parasitic values. The method presented here is fast and convenient, because it directly measures the most important parasitic components and is suitable for a frequency range of 100 MHz to about 2 GHz. At higher frequencies, this measurement technique would be limited by strays in the measurement setup and by the simple component models used.

The two basic assumptions of the measurement technique presented here are that capacitors of 2% tolerance value are available and that we can measure frequency accurately. Before we can use a 2% capacitor as an impedance standard, however, it is important to determine its series inductance by the method in Figure 3.1(a). Capacitor series inductance will create a resonance, whose frequency can be measured. The terms *tracking generator* and *spectrum analyzer* are used for illustration only; a network analyzer, or signal generator with a power sensor, can be used as well. The measurements are performed in a coaxial environment with all connections kept as short as possible.

In both measurements, the notch depth can be roughly correlated to the component Q. Once we have obtained an accurate capacitor model from Figure 3.1(a), then that capacitor can be used to resonate with an unknown coil to determine its apparent inductance (which will include the effect of its stray parallel capacitance).

Mutual inductance between two coils can be obtained from Figure 3.2 by resonating the primary with a capacitor near the frequency of interest. Measure the frequency shift when the secondary is short-circuited.

$$M = \sqrt{L_1 L_2 \left[1 - \left(\frac{f_2}{f_1} \right)^2 \right]}$$

$$k = \sqrt{1 - \left(\frac{f_2}{f_1} \right)^2}$$

M = mutual inductance between coils (H)

k = coefficient of coupling, $k < 1$

L_1 = primary inductance (H)

L_2 = secondary inductance (H)

$f_1 > f_2$

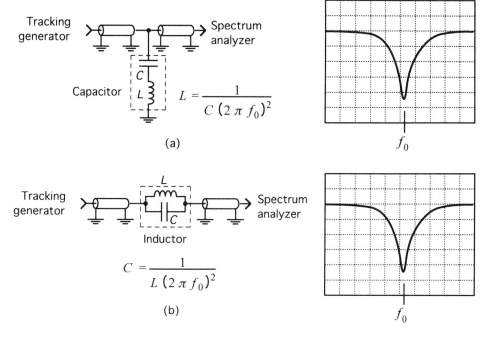

Figure 3.1 Measurement of capacitor and inductor parasitics: (a) Capacitor series inductance, (b) inductor parallel capacitance.

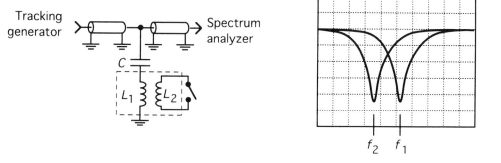

Figure 3.2 Measuring mutual inductance.

The winding polarity can be determined by connecting the primary and secondary in series. If the total inductance is greater than the sum of the individual inductances, then the polarity is series aiding. If the total inductance is less, then the polarity "dots" are at the junction of the windings.

A tapped coil is a special case of coupled inductors. The component values for a tapped coil can be measured with the help of Figure 3.3. *Tap* is defined as the ratio between the number of turns from the reference terminal to the tap connection and the total number of turns in the coil.

In general, $L \neq L_1 + L_2$, but $L = L_1 + L_2 + 2M$, and $k = [M^2/(L_1 L_2)]^{1/2}$. The following three measurements are required (adapted by permission from [1]):

1. First, measure inductance between nodes 2 and 3 with node 1 open. This gives L.
2. Next, measure inductance between nodes 2 and 3 with node 1 shorted to 3. This measurement yields $A = L_1 - \dfrac{M^2}{L_2}$
3. Finally, measure inductance between nodes 2 and 1 with node 3 open. This gives L_1.

The coefficient of coupling can then be computed as $k = \sqrt{1 - \dfrac{A}{L_1}}$, and

$$tap = \frac{100\sqrt{L_2}}{\sqrt{L_1} + \sqrt{L_2}}$$

3.2.2 Quality Factor

The Q of a resonant circuit can be measured by very lightly coupling the input to the magnetic field and the output to the electric field, with an insertion loss of greater than 20 dB at resonance. As shown in Figure 3.4, this can be done inductively with a small grounded loop and capacitively with an open-ended coax. Ideally, the input and output connections should not couple to each other by any means except through the resonator. Direct capacitive coupling between the input and the output of the measurement setup will produce an asymmetrical resonance curve whose peak may not be very prominent. If the input and the output are not loosely coupled to the resonator, their resistive impedance will distort the measurement.

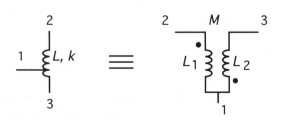

Figure 3.3 Tapped-coil equivalence.

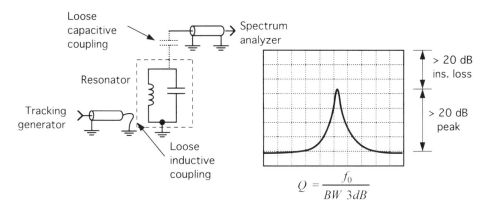

Figure 3.4 Measuring unloaded Q.

The measurement relies on accurate knowledge of the resonance amplitude shape and frequency, as given by (4.65).

The measurement of Figure 3.4 applies to transmission line resonators, as well as parallel connection of discrete capacitor and inductor. The overall Q of a discrete parallel LC circuit will be a combination of inductor and capacitor Q, according to the usual formula $\dfrac{1}{Q} = \dfrac{1}{Q_C} + \dfrac{1}{Q_L}$ which is applicable in this case because the inductor and capacitor reactance slopes are equal at resonance.

When measuring inductor Q, we often can assume the capacitor to be ideal, and the measurement of Figure 3.4 then yields inductor Q directly.

The measurement of capacitor Q requires a low-loss $\lambda/4$ coaxial cavity of known properties, which is then loaded with the capacitor under test at its open end. The resulting measurement of Q and resonant frequency shift can be used to derive the actual capacitor Q and capacitance value. The mathematical derivation of capacitor Q using this procedure is surprisingly complicated, mainly because the resonator's Q varies with frequency, and also because the familiar formula $\dfrac{1}{Q_{\text{Total}}} = \dfrac{1}{Q_{\text{Capacitor}}} + \dfrac{1}{Q_{\text{Resonator}}}$ *cannot* be used in this case, because the reactance slopes of the resonator and the capacitor are not equal at the frequency of interest. The general definition of Q from (4.55) must be used to calculate both the resonator Q at the lower frequency and the Q of the capacitor-resonator combination. The procedure and the theoretical background for calculation of the equivalent series resistance (ESR) are outlined next.

First, measure resonant frequency, f_0, and unloaded Q_0 of the coaxial resonator by itself, without any capacitor. The line attenuation, α, can be calculated from (4.64). Also note from (4.67) that for an air-filled coaxial resonator α varies as $\sqrt{f}$, so that α at any frequency can be determined once it is known at f_0.

Next, connect the capacitor under test between the open end of the resonator and ground and measure a new Q and resonant frequency, Q_M and f_M, respectively.

Now, using (4.55) with $N = M = 2$, the Q of the capacitor-resonator configuration is:

$$Q_M = \frac{1}{2}\frac{\left(X - \dfrac{1}{\omega_M C}\right) + \omega_M \dfrac{dX}{d\omega} + \dfrac{1}{\omega_M C}}{ESR + R} = \frac{1}{2}\frac{\omega_M \dfrac{dX}{d\omega} + \dfrac{1}{\omega_M C}}{ESR + R}$$

from which the ESR can be determined. Here the term $X - \dfrac{1}{\omega_M C}$ became zero due to resonance. Substituting expressions for transmission line X and R, available from (4.23), (4.25), (4.58), (4.59), and (4.60) and simplifying, we finally obtain:

$$ESR = \frac{1}{4}\frac{\pi^2 C Z_0 f_M^2 + f_0\cos^2\!\left(\dfrac{\pi f_M}{2 f_0}\right)}{\pi C Q_M f_0 f_M \cos^2\!\left(\dfrac{\pi f_M}{2 f_0}\right)} - \frac{Z_0\sinh\!\left(\dfrac{c\alpha_0\sqrt{f_M}}{2\sqrt{f_0^3}}\right)}{\cos\!\left(\dfrac{\pi f_M}{f_0}\right) + \cosh\!\left(\dfrac{c\alpha_0\sqrt{f_M}}{2\sqrt{f_0^3}}\right)}$$

where

$$C = \frac{1}{2\pi f_M Z_0 \tan\!\left(\dfrac{\pi f_M}{2 f_0}\right)}$$

Also,

$$ESR = \frac{1}{2\pi f_M C Q_C} = \frac{D}{2\pi f_M C} \tag{3.2}$$

Q_C = capacitor quality factor

Q_M = measured quality factor with capacitor loading a coaxial resonator at resonant frequency, f_M.

f_0 = resonant frequency of coaxial cavity without capacitor

f_M = resonant frequency of coaxial cavity with capacitor connected between open end and ground

$\omega_M = 2\pi f_M$

α_0 = line attenuation at f_0 (Nepers/m), calculated from Q_0 by (3.5)

c = speed of light, 2.9979×10^8 (m/s)

X = equivalent series reactance of transmission line

R = equivalent series resistance of transmission line

Q_0 = unloaded Q of coaxial resonator without capacitor at f_0

ESR = capacitor ESR (Ω)

C = capacitor value (F)

D = dissipation factor, loss tangent of capacitor dielectric material

For Q measurements of transmission lines, please consult Subsection 3.2.3; measurements of crystal Q are included in Subsection 2.14.1.

3.2.3 Single and Coupled Lines

Single transmission lines are described by their characteristic impedance, Z_0; physical length, l; velocity factor, v_p; and attenuation, α. Coupled transmission lines double the required number of values because even and odd modes of propagation have to be considered. The basic determination of characteristic impedance relies on measuring the input impedance when the line is open-circuited and when it is short-circuited. Since these impedances are reactive, the measurement can be performed in a similar manner to that described in Section 3.2, where we were measuring capacitors and inductors. While frequency is not an explicit parameter in these equations, it is assumed that all measurements are performed at the same frequency, unless stated otherwise.

$$Z_0 \approx \sqrt{|Z_{OC}||Z_{SC}|} \tag{3.3}$$

Z_0 = unknown transmission line characteristic impedance (Ω)

Z_{OC} = input impedance of line when open-circuited at the other end (Ω)

Z_{SC} = input impedance of line when short-circuited at the other end (Ω)

The line length must not be appreciably changed by the shorting operation, and we must avoid measuring the impedances at frequencies where the line is resonant (where line length is a multiple of $\lambda/4$). The measurement is approximate, unless stray capacitance of the open connection and stray inductance of the shorting connection are taken into account.

Velocity factor can be obtained by measuring the $\lambda/4$ resonant frequency of a length of transmission line, for example, a shunt open-circuited section.

$$v_p = \frac{1}{\sqrt{\epsilon_{eff}}} \approx \frac{4 f_0 l}{c} \tag{3.4}$$

v_p = velocity factor, dimensionless, $v_p \leq 1$

ϵ_{eff} = effective dielectric constant

f_0 = resonant frequency of $\lambda/4$ line (Hz)

l = physical length of line (m)

c = speed of light, 2.9979×10^8 (m/s)

While the attenuation of a line could be measured directly, there are errors introduced by mismatch; see (4.22). The best way to estimate line attenuation is to measure the unloaded Q of a $\lambda/4$ resonant section and invert (4.64) to obtain α from Q at the frequency of $\lambda/4$ resonance.

$$\alpha_N = \frac{\pi}{4Ql} \tag{3.5}$$

$$\alpha_{dB} = \frac{2.171\,\pi}{Ql} \tag{3.6}$$

α_N = line attenuation (Nepers/m)

α_{dB} = line attenuation (dB/m)

Q = measured unloaded Q of $\lambda/4$ resonant section

l = physical length of $\lambda/4$ transmission line (m)

When considering coupled-line parameters, there exists an important distinction between TEM coupled lines such as stripline, and non-TEM or quasi-TEM lines such as microstrip. The even- and odd-mode velocity factors are equal for TEM coupled lines, while for microstrip they generally are not. Any measurement of coupled line properties must take this distinction into account. Figure 3.5 shows the measurement setup for obtaining even and odd TEM characteristic impedances by treating each connection as a single transmission line (adapted from [1] by permission).

$$Z_{0e} = 2Z_a \tag{3.7}$$

$$Z_{0o} = \frac{Z_b^2}{Z_{0e}}$$

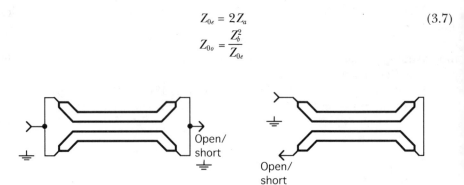

(a)   (b)

Figure 3.5 TEM mode coupled line measurement: (a) even mode, (b) odd mode.

Z_{0e} = even-mode characteristic impedance

Z_{0o} = odd-mode characteristic impedance

Z_a = characteristic impedance measured in Figure 3.5(a)

Z_b = characteristic impedance measured in Figure 3.5(b)

The even- and odd-mode velocity factors are equal for TEM structures and can be derived using (3.4) by treating the connection of Figure 3.5(a) as a single transmission line.

For non-TEM coupled lines the parameters Z_{0e}, ν_{pe}, and α_e can be derived using Figure 3.5(a) and (3.7). Figure 3.6(a) can be used to derive both ν_{pe} and ν_{po} in one measurement.

The frequency response plot of the circuit in Figure 3.6(a) will show two closely spaced notch frequencies; (3.4) can be used to calculate each mode velocity factor. Make use of the fact that ν_{pe} is normally less than ν_{po} to separate the two modes. The drawback of this technique is that a good short circuit may be difficult to provide. Reference [4] shows other approximate connections that can be used to avoid short circuits.

The odd-mode characteristic impedance cannot be obtained by the method in Figure 3.5(b) because the mode velocity factors are different. Use Figure 3.6(b) at a frequency well below that where the coupled sections are quarter-wave long.

$$Z_{0o} \approx 2Z_b - Z_{0e}$$

Z_{0o} = unknown odd-mode characteristic impedance

Z_b = characteristic impedance measured in Figure 3.6(b)

Z_{0e} = previously determined even-mode characteristic impedance

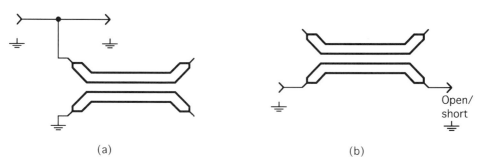

(a)   (b)

Figure 3.6 Non–TEM mode coupled line measurements: (a) velocity factor, (b) odd mode characteristic impedance.

All the transmission line properties mentioned in this section can also be obtained by computer methods using approximate numerical solution of Maxwell's equations for a cross-section of the physical structure [2,3] or from closed-form equations if available [4].

3.3 DIRECTIVITY AND RETURN LOSS

Return loss is a measure of how good an impedance match is present at a certain stage interface. A 20-dB return loss is considered very good, because it results in less than 0.1 dB mismatch loss. Precision devices may require measurement of return loss better than 20 dB. Accurate return loss measurements require surprisingly high directivity from the measurement setup. Figure 3.7 shows the range of return loss measurement errors when 20-dB and 30-dB directivity test setups are used. The shaded regions represent measurement uncertainty where the phase of the measured return loss signal interacts with the phase of an internal signal present due to limited directivity in the measurement setup.

In measurement of 15-dB return loss using a directional coupler or VSWR bridge with 20-dB directivity, the uncertainty in the measurement is from −4 dB to +7 dB. In other words, a true 15-dB return loss can produce a measured return

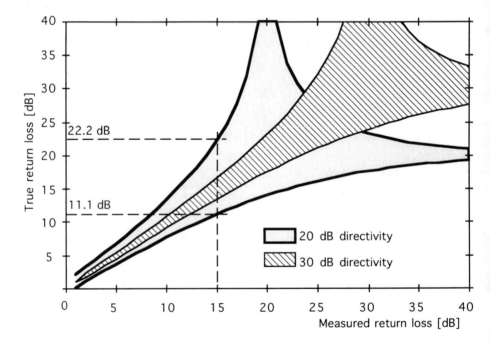

Figure 3.7 Relationship between return loss and directivity.

loss in the range between 11.1 dB and 22.2 dB, depending on the phase of the reflected signal relative to the incident signal, when using a 20-dB directivity coupler.

This error is surprisingly large and reaches a maximum when we try to measure return loss that is approximately equal to the directivity. Measurements made using low-directivity couplers can be deceiving. A measurement of 40-dB return loss using a 20-dB coupler really indicates that you are measuring a 20-dB return loss device. Trying to optimize return loss in such a measurement setup may be counterproductive, because an improved measured return loss does not necessarily mean improved true return loss.

$$RL'_{max} \approx -20 \log(10^{-RL/20} + 10^{-D/20}) \tag{3.8}$$

$$RL'_{min} \approx -20 \log|10^{-RL/20} - 10^{-D/20}| \tag{3.9}$$

RL'_{max} = upper limit of true return loss (dB)

RL'_{min} = lower limit of true return loss (dB)

RL = measured return loss (dB)

D = directivity of measurement setup (dB)

3.4 FREQUENCY DEVIATION

Frequency deviation is a closely controlled parameter in FM communication systems, because excessive frequency deviation will result in adjacent channel interference. Frequency deviation, expressed in ±kHz, describes the maximum instantaneous frequency difference between the transmitted signal and its nominal center frequency. Its measurement method relies on the fact that the Bessel functions that describe the modulation sidebands when a pure sine wave modulating signal is used cause the carrier component to disappear from the frequency spectrum for specific combinations of frequency deviation and modulating frequency. The most frequently used frequency component is the carrier, which disappears when the ratio between the frequency deviation and the modulating frequency is 2.405, which is the first zero crossing of Bessel function J_0.

The first sideband pair disappears when the modulation index (ratio between frequency deviation and modulating frequency) is 3.8; this is the first zero crossing of Bessel function J_1.

More accurately, the carrier disappears for modulation indices of 2.40483, 5.52008, 8.65373, 11.795. First sidebands disappear for modulation indices of 3.8317, 7.0156.

For example, to calibrate an FM source for 5.0-kHz peak deviation, use a sine wave audio modulating signal of 2.079 kHz and watch for the carrier spectral component to disappear from a spectrum analyzer display as the frequency deviation is adjusted.

The spectrum analyzer screen in Figure 3.8 corresponds to 2.405-kHz deviation of 1-kHz modulating frequency, or 5-kHz deviation of 2.079-kHz modulating frequency. Both represent a modulation index of 2.405.

3.5 FREQUENCY SYNTHESIZERS AND PHASE-LOCKED LOOPS

A typical phase-locked loop block diagram is shown in Figure 3.9.

When the loop is locked, $f_{out} = Nf_r$. The divider N is usually programmable. There may also be a frequency divider for the reference frequency input. The loop filter $F(s)$ is typically a low-pass filter having corner frequency f_L. The extra low-pass filter formed by R_{SL} and C_{SL}, and whose corner frequency is much higher than the main loop filter corner frequency, is required to attenuate high-frequency signals originating in the phase detector. Typically, the phase detector puts out very narrow,

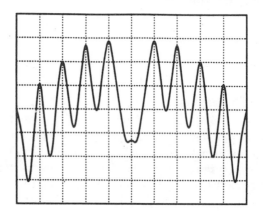

Figure 3.8 Spectrum analyzer display for first Bessel null.

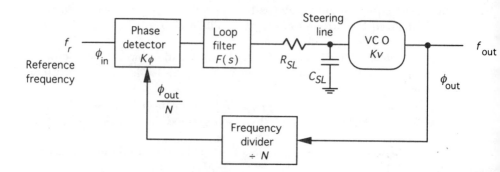

Figure 3.9 Block diagram of a typical phase locked loop.

fast pulses that are rich in harmonics; these harmonics are not sufficiently attenuated by lead-lag type of filters, and extra filtering is required for optimum loop performance. C_{SL} is also required to attenuate thermal noise originating in the varactor diode series resistor, as shown in Section 4.14 and (4.53).

The SSB phase noise of the output signal at f_{out} is determined by the VCO for frequency offsets from the carrier much greater than f_L. For frequency offset much less than f_L, the contributors to SSB phase noise are the SSB phase noise of the reference signal, phase detector noise, divider noise, and thermal noise if loop filter gain is large.

The closed-loop transfer function determines the lock time, or how long it takes for the frequency to stabilize, following a change in the divider ratio N. A frequency synthesizer typically operates in one of the following three modes:

- *Static conditions, loop is not locked.* The steering line is at its low or high voltage extreme, and the relationship between frequencies and phases is indeterminate.
- *Static conditions, loop is locked.* $f_{out} = Nf_r$, and there is a constant phase difference at the phase detector inputs.
- *Dynamic conditions, where typically the N has just changed, and the loop is in the lock acquisition mode.* This mode then approximates the step response of the system described by (2.55). To obtain the step response, integrate the inverse Fourier transform of (2.55); recall that the inverse Fourier transform yields the impulse response, whose time integral is the step response. This is only an approximation because the loop may enter a nonlinear mode of operation due to limited steering line voltage range, the phase detector gain may deviate from linear slope for large phase differences, and the phase detector may not have enough current drive capability to quickly charge or discharge loop filter capacitors.

Secondary effects in frequency synthesizer designs include the following:

- In wide-tuning VCOs, the steering line sensitivity, K_V, changes with RF frequency. Thus, all the synthesizer properties, including lock time, will also be affected.
- Delay in frequency divider [5] and phase detector introduces additional phase error, which needs to be taken into account in fast-tuning synthesizers with wide loop bandwidths.
- The reference frequency will modulate the VCO to some extent by leaking through the loop filter, entering the VCO by ground loops, or via the power supply leads. A high-performance frequency synthesizer will need careful layout, shielding, and bypassing, complete with separate voltage regulators and ground returns for the analog and digital portions of the circuit.
- Fast-locking synthesizers place challenging requirements on leakage from the steering line, loop capacitor dielectric absorption, phase detector–charge

pump imbalances, VCO post-tuning drift and thermal transients, VCO AGC performance, and load pulling. Steering line leakage always produces a transient frequency error in one direction following a frequency change, while dielectric absorption produces transient frequency error in both directions, depending on the previous steering line voltage.

- Phase detector second-order phenomena, such as a dead zone for small phase errors, may effectively open the loop near phase lock, introducing unpredictable transients.

Both K_ν and K_ϕ can be obtained by direct measurement on the VCO and phase detector, respectively. Introduce a small voltage change on the VCO steering line and observe the resulting frequency change. Usually the frequency responses of VCOs and phase detectors are much faster than the response of the other loop components; therefore, K_ν and K_ϕ may be assumed constant and independent of the modulating frequency.

$$K_\nu(s) \approx K_\nu \approx 2\pi \frac{\Delta f}{\Delta V}$$

K_ν = VCO conversion factor, steering line sensitivity (rad/s/V)

Δf = measured change in VCO frequency (Hz)

ΔV = small imposed change on steering line voltage (V)

The gain of a linear phase detector can be measured by using two slightly different frequencies at its inputs, which produce a beat note at its output, whose peak-to-peak time domain voltage swing is therefore produced by a phase change of π radians.

$$K_\phi \approx \frac{\Delta V_{p\text{-}p}}{\pi}$$

K_ϕ = gain of linear phase detector (V/rad)

$\Delta V_{p\text{-}p}$ = peak-to-peak voltage at phase detector output when input frequencies are slightly different. (V)

Measurement of lock time requires a fast frequency counter, frequency domain modulation analyzer, or any of the circuits of Subsection 2.7.2. One possible method using an FM modulation analyzer with dc coupled demodulated FM output is shown in Figure 3.10. Tune the modulation analyzer to expect the destination frequency and then program the synthesizer to switch from the source to the destination frequency. The modulation analyzer output will be a time function of the frequency

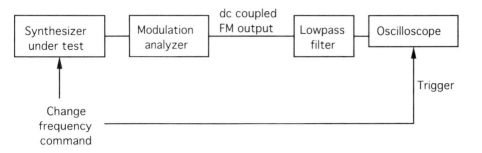

Figure 3.10 Measuring synthesizer lock time.

difference between the expected destination frequency and the actual synthesizer output frequency. Care must be taken to disable autoranging and AGC functions of the modulation analyzer, which could cause the dc FM output to saturate, producing its own recovery transient, possibly masking the true performance of the device under test.

Additional information on frequency synthesis techniques can be found in [6] and [7].

3.6 INTERCEPT POINT

The background theory of intercept points as measures of circuit and system linearity was explained in Subsection 1.1.6. This section deals with the reliable measurement of intercept points of individual circuit stages. These measurements can then be combined into a system calculation, as mentioned in Subsection 1.1.6.

Highly linear devices with high intercept points have to be measured carefully to ensure that the IM distortion is not produced in the measurement instruments. The test setup must always be checked to make sure that the IM signals do not originate in the spectrum analyzer used for the measurement. Another possibility for generating IM in the test setup is when the various signal generators are not sufficiently isolated from each other and produce IM in their output sections. Ferrite power combiners can also produce IM at sufficiently high input levels. Use circulators or attenuators at the signal generators' outputs and make sure that the power combiner is not mismatched at its output by the device under test. Use an attenuator at the power combiner's output to ensure best isolation for its inputs.

During measurement of third-order IM with two equal amplitude input signals and when the two distortion products are not equal in amplitude (similar to Figure 3.20), there are two or more contributors to IM. While it is possible that the device under test produces IM by more than one mechanism, the more likely situation is that the measurement setup is contributing its own IM and distorting the measurement results.

3.6.1 Second-Order Intercept Point (IP2)

Measurement of mixer IP2 is shown in Figure 3.11 and Figure 3.12. Attenuators may be placed at the mixer ports during broadband measurements.

$$IP2 = A + \Delta \tag{3.10}$$

IP2 = input second-order intercept point (dBm)

A = level of RF input signal at mixer RF port (dBm)

Δ = difference between desired IF signal and its second harmonic (dB)

Normally, frequency f in Figure 3.12 would be the IF frequency, and $2f$ would be its second harmonic. In many cases, the IF port of the mixer is accessible only

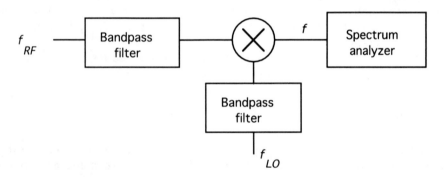

Figure 3.11 Mixer IP2 measurement.

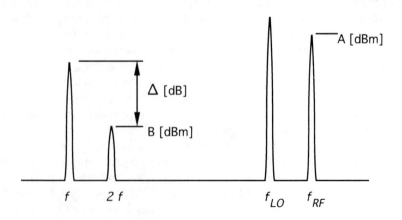

Figure 3.12 Frequency and amplitude relationships for mixer IP2 measurement.

through a bandpass filter, and the two signals for measuring Δ cannot be displayed at the same time. The preferred measurement method, then, would be a two-step process, with measurements performed at the IF frequency only (assume low-side injection):

    Step 1: $f_{RF} = f_{LO} + f_{IF}$, measure conversion loss of mixer

    Step 2: Shift the RF frequency to the 1/2 IF spur: $f_{RF} = f_{LO} + 0.5 f_{IF}$, measure B, level of $2f$ signal, which is now at the IF frequency.

$$\Delta = A - B - CL$$

Δ = parameter for use in IP2 equation (3.10) (dB)

A = input RF level (dBm)

B = output level of spurious response (dBm)

CL = conversion loss of mixer (dB)

    Bandpass filters in the measurement setup are required to remove the second harmonics of f_{RF} and f_{LO}, which would adversely interfere with the measurement.

    Measurement of IP2 for a two-port device such as an RF amplifier is shown in Figure 3.13 and Figure 3.14.

$$IP2 = A + \Delta \qquad (3.11)$$

IP2 = input second-order intercept point (dBm)

A = level of RF input signals at amplifier input (dBm)

Δ = difference between output signal level and second-order distortion
      products (dB)

G = device gain (dB)

    It is always advisable to verify the capability of the measurement setup by connecting the spectrum analyzer directly to the output of the bandpass filter and making sure that the level of distortion products generated by the setup is much lower than during measurement of the actual device under test.

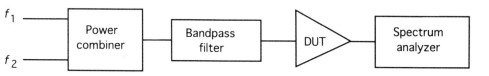

Figure 3.13 Amplifier IP2 measurement.

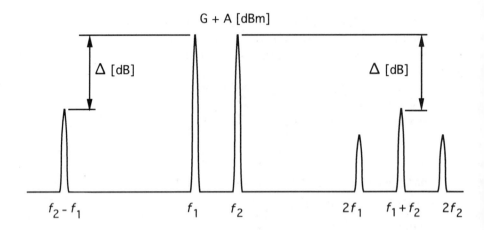

Figure 3.14 Spectrum analyzer display for amplifier IP2 measurement.

3.6.2 Third-Order Intercept Point (IP3)

Measurement of mixer IP3 is illustrated in Figure 3.15 and Figure 3.16. High-performance mixer measurements may require the addition of attenuators at all mixer ports to avoid reflection of harmonics from the test setup back into the mixer. On the other hand, it may be desirable to duplicate the same impedance terminations at mixer ports as would be present in the intended application, because a device's IP3 may depend very strongly on its terminating impedances.

$$IP3 = A + \Delta/2 \tag{3.12}$$

IP3 = input third-order intercept point (dBm)

A = level of RF input signals at mixer RF port (dBm)

Δ = difference between desired IF signal and IM products (dB)

IP3 measurement for a two-port device is shown in Figure 3.17 and Figure 3.18.

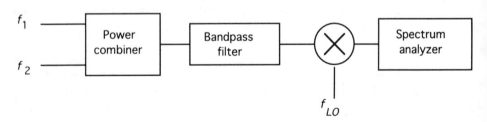

Figure 3.15 Measurement of mixer IP3.

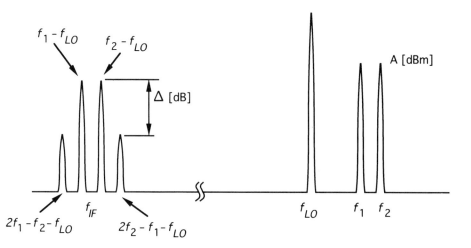

Figure 3.16 Frequency and amplitude relationships for mixer IP3 measurement.

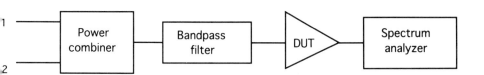

Figure 3.17 Amplifier IP3 measurement.

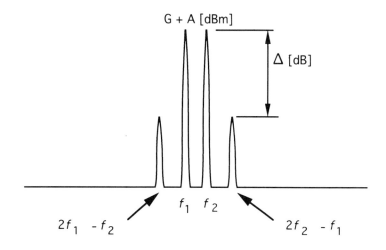

Figure 3.18 Spectrum analyzer display for amplifier IP3 measurement.

$$IP3 = A + \Delta/2 \tag{3.13}$$

IP3 = input third-order intercept point (dBm)

A = level of RF input signals at amplifier input (dBm)

Δ = difference between desired output signal and IM products (dB)

G = device gain (dB)

Connect the spectrum analyzer directly to the output of the bandpass filter and make sure that the level of distortion products generated by the setup is much lower than during the actual measurement.

3.6.3 nth-Order Intercept Point

The usefulness and measurement of higher-order intercept point is best illustrated by an example similar to the one in Subsection 1.1.6.3. A certain receiver with an IF of 73.35 MHz is designed to receive a signal at 442 MHz. This receiver will be susceptible to an undesired (4,5) spurious response (refer to Subsection 1.1.4 for notation) at 442.475 MHz, because $5f_{LO} - 4f_{RF} = 5(368.65) - 4(442.475) = 73.35$ MHz. This is a fourth-order process and can be analyzed by measuring the level of 73.35-MHz signal at the mixer's output when a certain level of 442.475-MHz signal is fed into the mixer's RF port, while keeping the LO frequency the same as that required for reception of a signal at 442 MHz ($f_{LO} = 368.65$ MHz).

Assume that the output level of the 73.35-MHz signal is −65 dBm when the input 442.475-MHz signal is at −10 dBm. Then by use of (1.19) the fourth-order intercept point is +6.33 dBm, assuming a 6-dB conversion loss in the mixer. Figure 3.19 shows the amplitude and frequency relationships.

Recognizing that the levels used in this measurement will not be encountered by the receiver in actual use, we can calculate the spurious response amplitude at any other input level by recalculating Δ in (1.19) using the new input level and an IP4 of +6.33. Thus, an input level of −23.75 dBm will cause an output level of −120 dBm, giving less than a 100-dB spurious response rejection assuming the receiver sensitivity is near −120 dBm.

The reason why this (4.5) spurious response is considered to be a fourth order product, rather than a ninth order product, is that the LO amplitude is constant and therefore its multiplier does not enter into the calculation, as explained in Subsection 1.1.6.3.

3.7 INTERMODULATION DISTORTION

The following fast troubleshooting technique can be used to determine which stage in a cascade of devices is primarily responsible for the observed intermodulation

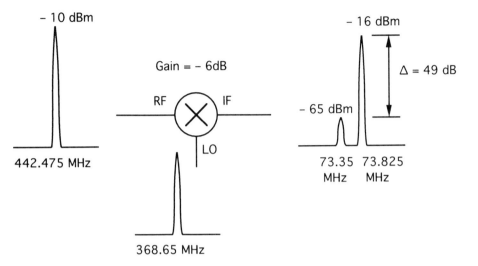

Figure 3.19 Example of fourth-order intercept point.

Introduce a small, controlled amount of attenuation (such as a large shunt resistor) between two stages, while monitoring the output signal and IM distortion products on a spectrum analyzer. If the signal and IM products drop in level by the same amount, the offending stage is before the node where the resistor was touched. If the IM products decrease in level more than the desired signal, then the offending stage is after the node with the external resistor. Then move the resistor up or down the chain, observing these effects, until the stage-generating IM distortion is isolated.

In Section 3.6 we considered IM interference resulting from two input signals of equal amplitude. The situation where two input signals of different amplitudes are encountered cannot be reduced to an equivalent equal amplitude case, because the two IM sidebands will be of different amplitudes, as shown in Figure 3.20.

$$\Delta = 2\text{IP3} - A_1 - A_2$$

Δ = difference between IM product and adjacent linear signal (dB)

G = stage gain (dB)

IP3 = input intercept point (dBm)

A_1 = lower-frequency input signal level (dBm)

A_2 = higher-frequency input signal level (dBm)

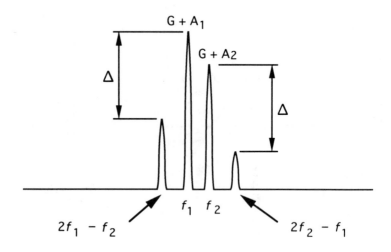

Figure 3.20 Output IM spectrum with input signals of different amplitudes.

3.8 MIXER NOISE BALANCE

Mixer noise balance refers to the tendency of mixers to transfer some amount of local oscillator AM noise to the IF port. This reduces the output S/N ratio by a larger amount than would be expected just from the mixer conversion loss. This becomes important in receiver sensitivity calculations (see Subsection 1.1.2). Double-balanced mixers typically have 30 dB of noise balance, while single-ended mixers do not have any LO AM noise rejection at the image frequency, requiring a separate injection filter to suppress this noise.

The transfer of local oscillator noise to the mixer output happens at the LO fundamental as well as the LO harmonics. Noise contribution from the harmonics is often more important than contribution from the fundamental. Figure 3.21 illustrates the frequency relationships.

The key concept is that any signal, including noise, contained in the LO spectrum at frequencies $nf_{LO} \pm f_{IF}$ will be transferred to the mixer's IF port.

Usually the noise level is far above thermal due to LO buffer amplifiers; the mixer takes the noise at $\pm f_{IF}$ from the fundamental and harmonics and transfers it to the IF port with some amount of attenuation.

The measurement of mixer noise balance uses a low-level, discrete signal at the appropriate $nf_{LO} \pm f_{IF}$ frequency (one at a time), instead of noise, as shown in Figure 3.22; the LO signal remains fixed at its operating frequency, but the low-level sideband signal changes frequency depending on which $nf_{LO} - f_{IF}$ sideband we want to measure. The level of the sideband signal should be much lower than the LO signal level.

$$M_S = A_S - A_{IF}$$

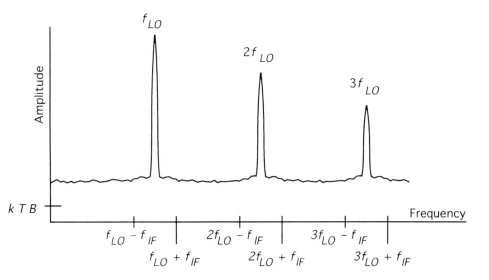

Figure 3.21 Local oscillator output spectrum including noise.

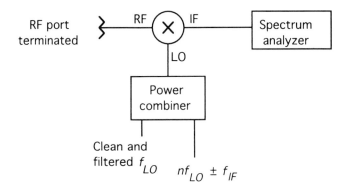

Figure 3.22 Measurement of mixer noise balance.

M_S = mixer noise balance at nth LO harmonic (dB)

A_{IF} = level measured on spectrum analyzer at f_{IF}. (dBm)

A_S = input level of $(nf_{LO} \pm f_{IF})$ signal at mixer LO port (after combiner) (dBm)

Each LO harmonic has two noise sidebands, we need $2n$ mixer noise balance measurements to characterize a mixer up to the nth LO harmonic. Mixer noise balance numbers are important contributors to receiver sensitivity degradation which can be quantified by (1.8).

3.9 NOISE TEMPERATURE AND NOISE FIGURE

Noise temperature and noise figure are measures of undesirable noise power introduced into a signal path. These measures are important because noise ultimately limits the S/N ratio. To obtain a desired S/N ratio, the signal power has to be above noise power by the required amount. The S/N ratio can be improved by increasing the signal power, up to a level at which IM products are produced at a level comparable to the ambient noise level. Analysis of noise levels is particularly important in calculations of receiver sensitivity. Additional noise introduced by circuit stages is almost always undesirable; the only exception is in design of a noise source.

Effective noise temperature of a circuit stage is most conveniently measured by means of the Y-factor method, where Y refers to the ratio of two noise powers. A calibrated noise source is required for this measurement; it can be a noise source, or it can be a 50-Ω termination at a known elevated or depressed temperature. For the purpose of this example, we assume that one of the two required noise sources is a 50-Ω load at room temperature, while the other one is a calibrated noise source with known excess noise ratio (ENR), which is the decibel noise power above thermal. The noise figure can then be derived from the effective noise temperature by a simple equation.

Noise power N_1 is first measured with the noise source connected; noise power N_2 is measured with a 50-Ω input termination, as shown in Figure 3.23. The Y-factor is then the ratio of these powers.

$$T_e = \frac{T_0(10^{ENR/10} - Y_1)}{Y_1 - 1} - \frac{T_{em}}{G} \tag{3.14}$$

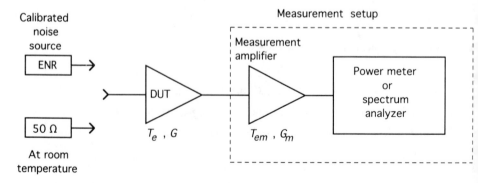

Figure 3.23 Measurement of effective noise temperature.

T_e = effective noise temperature of device under test (K)

T_0 = 290 K

T_{em} = effective noise temperature of test setup (K)

Y_1 = Y-factor ratio of two noise powers, N_1/N_2 (linear)

ENR = excess noise ratio of noise source (dB)

G = gain of device under test (linear)

N_1 = measured noise power with noise source connected (W)

N_2 = measured noise power with 50-Ω input termination (W)

This measurement requires that the device gain, as well as the measurement setup noise figure be known. While DUT gain is easily obtained, the test setup noise figure may not be known. In such a case, the DUT and measurement amplifier are exchanged and another similar Y-factor, Y_2 is obtained. Using Y_2 as well as Y_1, we can derive the DUT effective noise temperature using the two known amplifiers' gains:

$$T_e = \frac{G_m T_0}{(GG_m - 1)}\left[\frac{(10^{\text{ENR}/10} - Y_1)G}{Y_1 - 1} - \frac{10^{\text{ENR}/10} - Y_2}{Y_2 - 1}\right] \qquad (3.15)$$

T_e = effective noise temperature of device under test (K)

T_0 = 290 K

Y_1 = Y-factor, N_1/N_2 when DUT is the first device (linear)

Y_2 = Y-factor, N_1/N_2 when measurement amplifier is the first device

(i.e., DUT and measurement amplifier are exchanged) (linear)

ENR = excess noise ratio of noise source (dB)

G = gain of device under test (linear)

G_m = gain of measurement amplifier (linear)

N_1 = measured noise power with noise source connected (W)

N_2 = measured noise power with 50-Ω input termination (W)

Device noise figure can be obtained from the effective noise temperature by using (3.16).

$$F = 10\,\log\left(1 + \frac{T_e}{T_0}\right) \qquad (3.16)$$

F = noise figure (dB)

T_e = effective noise temperature (K)

T_0 = 290 K

A practical measurement setup will always require the measurement amplifier because the power meter or spectrum analyzer will not have enough sensitivity to detect low-level amplified thermal noise without error. The gain of the measurement amplifier is not important, as long as it is high enough to produce meaningful indication on the power meter. As long as the noise source ENR is greater than about 15 dB, the uncertainty in the ENR and the uncertainty in measurement of Y each result in approximately decibel-for-decibel uncertainty in measurement of device noise figure.

It is theoretically possible to obtain the ENR of an unknown noise source by performing the three measurements shown in Figure 3.24:

1. Measure output noise power with noise source connected at input, N_1.

2. Measure output noise power when noise source is attenuated by some known amount of attenuation, N_2.

3. Measure output noise power when input is terminated in 50 Ω at 290 K, N_3.

$$\text{ENR} = 10 \log\left(\frac{Y_2 - Y_1}{Y_2 + \dfrac{-Y_1 - 10^{A/10} + 1}{10^{A/10}}}\right) \tag{3.17}$$

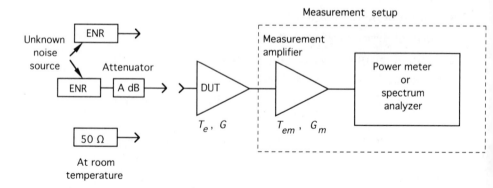

Figure 3.24 Approximate determination of unknown noise source ENR.

$Y_1 = N_1/N_3$ (linear)

$Y_2 = N_2/N_3$ (linear)

N_1 = output noise power with noise source connected (W)

N_2 = output noise power with noise source and attenuator (W)

N_3 = output noise power when input is terminated in 50 Ω at 290 K (W)

A = attenuator loss (dB)

The reason this method is approximate is that the denominator of (3.17) contains the difference between two nearly equal numbers, neither one of which can be obtained with great accuracy. Contrary to intuition, the best results are obtained with high T_e in the measurement setup, low ENR, and attenuator loss approximately equal to or greater than half of ENR.

For best accuracy, the amount of impedance mismatch introduced into any noise measurement test setup must be kept to a minimum; this is especially important when making different connections at the DUT input. At least 20-dB return loss should be maintained.

A spectrum analyzer can be conveniently used for making noise *ratio* measurements, but in measurement of *absolute* noise power there are three sources of error in a conventional spectrum analyzer. First, spectrum analyzers use a peak detector for amplitude measurements, and their display is calibrated to read correctly for rms sine wave signals. Noise power requires true rms measurement. The correction factor is about 2.5 dB; true noise power is 2.5 dB higher than displayed on a spectrum analyzer with a peak detector. Second, the IF or resolution bandwidth is typically less than the equivalent noise bandwidth. This correction is usually less than 1 dB. Third, the display is logarithmic, which tends to overemphasize low readings.

As a result, the displayed noise (even with video averaging turned on) is 2 to 6 dB lower than the true rms noise level, depending on the spectrum analyzer. Whenever you need to measure absolute noise power, such as oscillator wideband noise, the measurement needs to be carefully calibrated. Do this with a noise source of known ENR, using Figure 3.25 and (3.18).

Set up the spectrum analyzer using the settings appropriate for your measurement and then measure the noise source. The displayed reading will be N_{cal}, in dBm. Then connect the unknown device, whose noise power output you want to determine, and measure N_{DUT}, in dBm. Both these noise measurements will be subject to the three errors mentioned, but we can nevertheless extract the true noise power of the DUT from the following equation:

$$T_{DUT} = 290 \times [10^{(N_{DUT} - N_{cal})/10}] \times (10^{ENR/10} + 1) + [10^{(N_{DUT} - N_{cal})/10} - 1]\left(T_{e1} + \frac{T_{e2}}{G_1}\right)$$

$$(3.18)$$

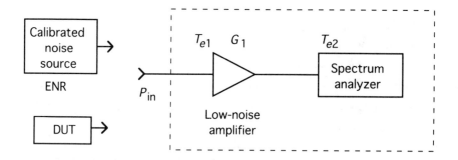

Figure 3.25 Calibrating absolute noise power measurement.

T_{DUT} = true equivalent noise temperature of measured device (K)

N_{cal} = power reading with calibrated noise source connected (dBm)

N_{DUT} = power reading with DUT connected (dBm)

ENR = excess noise ratio of calibrated noise source (dB)

T_{e1} = equivalent input noise temperature of low-noise amplifier (K)

T_{e2} = equivalent input noise temperature of spectrum analyzer (K)

G_1 = gain of low-noise amplifier (linear)

k = Boltzmann's constant, 1.38×10^{-23} J/K

The true noise power density of the DUT is then kT_{DUT}, in watts per hertz. We have obtained this number without having to know the properties of the spectrum analyzer detector or its equivalent noise bandwidth. If ENR is large, low-noise amplifier gain G_1 is large, and its noise temperature T_{e1} low, then (3.18) reduces to (3.19), which says that the true DUT noise power is above the calibrated noise source ENR by the same amount as the difference in decibels between measured N_{DUT} and N_{cal}.

$$N_0 \approx N_{DUT} - N_{cal} + ENR - 174 \qquad (3.19)$$

N_0 is the true noise power output of the device under test; it is also equal to $30 + 10 \log(kT_{DUT})$, in units of dBm per hertz.

This method can be used to obtain the ENR of an unknown source, if another calibrated noise source is available, in contrast to the method in Figure 3.24, where only an accurate attenuator is required.

3.10 OSCILLATOR OUTPUT IMPEDANCE, LOADED Q

The knowledge of an oscillator's output impedance can be used to estimate the amplitude flatness over frequency when connected to a certain load. It can also be used to control the noise introduced by the buffer stages. Loaded Q will contribute to SSB phase noise, and its knowledge is an important design tool for designing low-noise oscillators.

Oscillator output impedance can be measured by the load pull method in Figure 3.26, provided that the oscillator was designed to supply power into a resistive load. The load pull method will fail if the oscillator output impedance is reactive.

The measured power will vary as the length of adjustable line is changed [8,9]. Assuming that $Z_0 = 50\ \Omega$ and there is enough travel in the variable-length line to go around the Smith Chart, then $|\Gamma_2| = 0.3333$ (i.e., 25-Ω load giving VSWR = 2) and the measurement procedure is as follows:

1. Obtain the difference in level ΔM, in decibels, between maximum and minimum output power as the line length is changed.
2. Calculate $|\Gamma_1|$ from the following equation:

$$|\Gamma_1| = \frac{10^{\Delta M/20} - 1}{|\Gamma_2|(10^{\Delta M/20} + 1)} \tag{3.20}$$

$|\Gamma_1|$ = magnitude of oscillator reflection coefficient (linear)

ΔM = difference between maximum and minimum power reading (dB)

$|\Gamma_2|$ = reflection coefficient magnitude of test setup, theoretically 0.333

3. Measure the phase angle, α, of Γ_2 when line adjusted for maximum power meter reading.
4. Oscillator reflection coefficient is then $|\Gamma_1|$ at $-\alpha°$, and

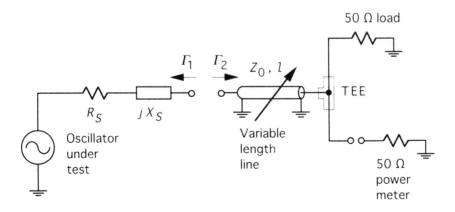

Figure 3.26 Measurement of oscillator output impedance.

$$R_S = \text{Re}\left[Z_0 \frac{1 + \Gamma_1}{1 - \Gamma_1}\right] \qquad (3.21)$$

$$X_S = \text{Im}\left[Z_0 \frac{1 + \Gamma_1}{1 - \Gamma_1}\right] \qquad (3.22)$$

R_S = oscillator source resistance, in series with X_S (Ω)

X_S = oscillator source reactance, in series with R_S (see Figure 3.26)

The impedance relationships are illustrated graphically on the Smith Chart in Figure 3.27.

Intuitively, if there is no power variation as the line length is changed, then the oscillator's output impedance is 50 Ω real.

If we use a spectrum analyzer instead of a power meter as the detector in this test setup, the oscillator's frequency will be observed to change as the line length is changed. This undesirable effect is called *load pull* and can be minimized by adding buffer amplifiers with low s_{12} to the oscillator's output.

The load pull measurement, when performed on an oscillator without any buffer amplifier connected, can supply an estimate of the oscillator's loaded Q. To pursue this idea further, let us develop a simple oscillator equivalent circuit. To emphasize again, the buffer amplifier must not be connected to the oscillator, and the oscillator's output impedance must not be very reactive.

In Figure 3.28, the main resonator is the parallel combination of known L and C, amplifier with gain G and coupling k to L provides the output signal, which is coupled to the external world through some unknown impedance Z and ideal transformer with unknown turns ratio n.

If we include the active device loading in R_p and take Z to the other side of the output coupling transformer, we end up with the circuit in Figure 3.29.

The output impedance Z_S of this circuit is known from the measurement illustrated in Figure 3.26 and relies on the fact that at resonance, the impedance on the left side of transformer is real and equal to R_{int}. Then $R_S = R_{int}/n^2$.

The quantities L and C are known in a typical oscillator design, while R_{int}, X_S and n are unknown. In our case, we managed to measure or derive X_S and can use that fact to determine n, the turns ratio of ideal output coupling transformer. Consider what happens to the resonant frequency when the output node A is left open and then is shorted to ground. When node A is open, the resonant frequency is essentially determined by L and C. When node A is shorted to ground, we introduce a reactance of $jn^2 X_S$ across the resonator. If the frequency goes up, X_S is positive (inductive); if the frequency goes down, X_S is negative (capacitive). Make sure that the signal pickup is very loosely coupled to the oscillator, so that the frequency measurement technique itself does not affect the frequency.

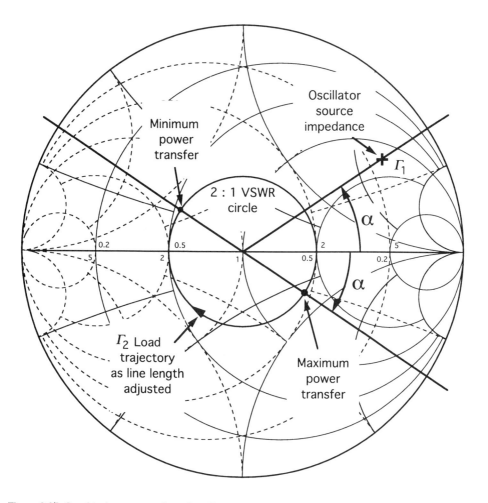

Figure 3.27 Graphical representation of oscillator impedance relationships.

To summarize, our objective is to obtain the values of R_{int} and n in the equivalent circuit in Figure 3.29 and from there obtain the loaded Q of the operating oscillator.

1. Measure oscillator output impedance using the load pull method in Figure 3.26. and (3.20), (3.21), and (3.22). This yields R_S and X_S.
2. The oscillator frequency shift with output open and shorted to ground supplies an estimate for n, the output coupling transformer turns ratio.

$$n = \sqrt{\frac{f_S}{2\pi(f_S^2 - f_0^2)\, CX_S}} \tag{3.23}$$

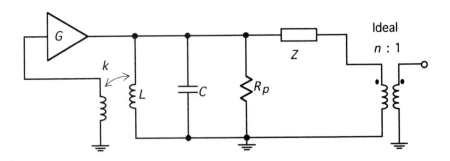

Figure 3.28 One possible oscillator equivalent circuit.

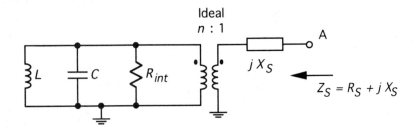

Figure 3.29 Simplified oscillator equivalent circuit.

n = turns ratio of transformer in equivalent circuit

f_s = oscillator frequency when output shorted (Hz)

f_0 = oscillator frequency when output open (Hz)

C = resonant circuit capacitance (F)

X_s = oscillator equivalent output reactance (Ω)

Note that (3.23) ensures that the quantity under the square root is always positive: When X_s is negative (capacitive), frequency shifts down, $f_s < f_0$, and vice versa.

3. $R_{int} = n^2 R_S$ and $Q_L = R_{int}/(\omega L)$ (3.24)

This procedure yields all component values in the equivalent circuit.

For example, let $C = 10$ pF, $f_0 = 505$ MHz when A is open, and $f_s = 500$ MHz when A is shorted to ground. A previous load pull measurement gave $R_S + j X_S = 30 - j20\ \Omega$.

Now, to pull the frequency from 505 MHz down to 500 MHz, we need an additional capacitance of $C' = 10$ pF$[(505/500)^2 - 1] = 0.201$ pF in parallel with our 10-pF capacitor, which represents a reactance of $-j1584\ \Omega$ at 500 MHz. This quantity is equal to jn^2X_S, as determined above, and can be used to determine $n = (1584/20)^{0.5} = 8.9$. Also, $R_{int} = n^2R_S = 2376$ Ω. As a result of this procedure, we obtain the equivalent circuit in Figure 3.30 for our oscillator. Its loaded Q is approximately $R_{int}/(\omega L) = 75$.

This equivalent circuit is useful in calculating the oscillator's SSB phase noise via Leeson's equation [10]. Once the buffer amplifier is connected to node A, the oscillator's new loaded Q (including loading of buffer stage) can be very easily obtained by computer simulation using the schematic in Figure 3.31. The loaded Q is a parameter in Leeson's equation and is notoriously difficult to obtain by direct measurement on an operating oscillator.

The frequency response of this circuit is a classical resonance curve (see Figure 4.15) and can be used to obtain the loaded Q.

$$Q_L = (\text{resonant frequency})/(\text{3-dB bandwidth})$$

If the buffer is so tightly coupled to the oscillator that the oscillator fails to oscillate when node A is shorted to ground, then introduce a small known impedance in series with A and track all the calculations as if it were in series with X_S. This procedure may also be necessary if there is no frequency shift when A is grounded—this is the lucky coincidence where $X_S = 0$.

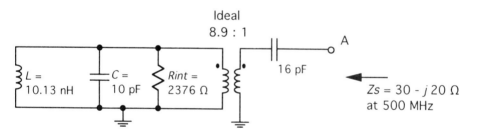

Figure 3.30 Example of complete equivalent circuit of 500-MHz oscillator.

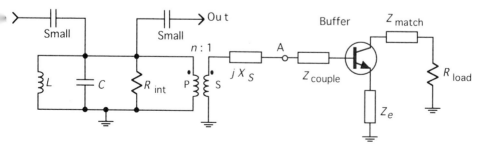

Figure 3.31 Oscillator loaded Q obtained by computer simulation.

Oscillator loaded Q can also be estimated if we can ascertain the active device's operating transconductance gain, either by simulation (harmonic-balance, or time-domain Spice) or measurement (dc bias current, AGC state). This concept relies on the fact that as losses are introduced into the circuit, the active device's transconductance must increase to keep loop gain at 1 in order for the circuit to keep oscillating. Simply keep adding resistive loss at a convenient circuit node until the gain increases by 3 dB. This means that the circuit's internal losses are now equal to the added losses. Since the latter is known, the former can be modeled by an equivalent resistance equal to the one just added. This is simply a consequence of the conservation of energy and is applicable to power increases of less than 3 dB, if that seems too extreme a change in operating point.

A completely different method of oscillator analysis relies on the fact that if the feedback loop can be broken at a convenient location, then the open-loop transfer function can be analyzed for voltage gain and phase relationships, which then yield the oscillation frequency and loaded Q. There are two basic difficulties with this technique: operation of the device is assumed linear, and once the loop is broken, the correct source and load impedances are unknown, and are therefore arbitrarily assumed resistive [11]. This second difficulty can be overcome when we realize that just before the loop is broken, the load impedance for the circuit is its own input impedance, as shown in Figure 3.32.

Using (4.78) and (4.79) we can derive the impedance conditions, which must be present when the loop is closed. The sign of the square root quantity is chosen such that $\Gamma_L = s_{11}$ when $s_{12} = 0$, and $\Gamma_S = s_{22}$ when $s_{12} = 0$.

$$\Gamma_L = \Gamma_{\mathrm{IN}} = \frac{1 - s_{21}s_{12} + s_{22}s_{11} \pm \sqrt{-4s_{22}s_{11} + (s_{21}s_{12} - s_{22}s_{11} - 1)^2}}{2s_{22}} \tag{3.25}$$

$$\Gamma_S = \Gamma_{\mathrm{OUT}} = \frac{1 - s_{21}s_{12} + s_{22}s_{11} \pm \sqrt{-4s_{22}s_{11} + (s_{21}s_{12} - s_{22}s_{11} - 1)^2}}{2s_{11}} \tag{3.26}$$

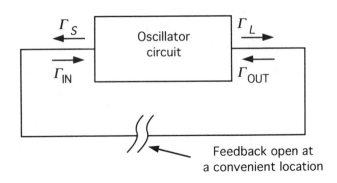

Figure 3.32 Oscillator analysis by breaking the feedback loop.

Γ_L = required load reflection coefficient for open-loop analysis

Γ_S = required source reflection coefficient for open-loop analysis

Γ_{IN} = input reflection coefficient of open-loop network, which is also terminated in Γ_{IN} at its output

Γ_{OUT} = output reflection coefficient of open-loop network, which is also terminated in Γ_{OUT} at its input

s_{ij} = s-parameters of open-loop network obtained by measurement or simulation

Thus, we can analyze the open-loop circuit by ensuring that the source impedance Γ_S is made equal to Γ_{OUT}, and load impedance Γ_L is set to Γ_{IN}. The open-loop network is first analyzed to obtain its s-parameters. Then (3.25) and (3.26) are used to calculate the proper source and load impedances to duplicate the conditions present when the loop is closed. The analysis is then repeated with the correct loading impedances to obtain the voltage gain, noise figure, phase, and group delay required for complete oscillator analysis. The circuit will oscillate at a frequency where the phase of the open-loop transfer function crosses zero degrees, as long as the open-loop gain is greater than 1. The phase zero crossing frequency does not necessarily coincide with a gain maximum.

The difficulty associated with assumed linear operation can be overcome by repeating the analysis with high-level s-parameters, if available. The loaded Q is related to the phase slope (group delay) *at the phase zero crossing frequency*.

$$Q_L = \omega \frac{\tau}{2} = -\frac{\pi f_0}{360} \frac{d\phi}{df} \qquad (3.27)$$

Q_L = oscillator loaded quality factor

ω = radian frequency, = $2\pi f_0$ (rad/s)

τ = group delay (s)

f_0 = frequency (Hz)

ϕ = phase of voltage gain transfer function (degrees)

The advantages of using computer simulation for analyzing oscillators are numerous. The effect of different buffer configurations on loaded Q and thereby on SSB phase noise can be established, statistical properties of Q_L can be obtained given the component value tolerances, and peak RF voltage across components (especially important for varactors) can be easily estimated. The power generated by oscillator can also be determined, and linear operation of buffer can be ensured. If buffer operation is nonlinear, SSB phase noise will be degraded due to AM to

PM conversion. AM noise contributed by the amplifier and power supplies can be much higher than PM noise. Conversion from one to the other by a nonlinear buffer should be avoided.

3.11 OSCILLATOR PHASE NOISE

Measurement of SSB phase noise is usually performed by the phase detector method, which is suitable for measurements at frequency offset greater than 100 Hz from carrier [12]. Figure 3.33 shows that the signal from the device under test is mixed with a low phase noise signal in phase quadrature. The lower sideband of the mixing process is at audio. The down-converted phase noise spectrum is amplified and examined on an audio spectrum analyzer. Quadrature between the two signals is maintained by feeding the baseband signal into the dc FM input of the low phase noise signal generator.

    The signals must be kept in phase quadrature producing zero dc at the phase detector output. This can be monitored using an oscilloscope. Since a dynamic range greater than 100 dB is required for many measurements, a low-noise audio amplifier must be used, followed by a switchable attenuator to increase the dynamic range at the high-signal level present during calibration. The raw measurement of noise on the audio spectrum analyzer must be converted to a single-sideband dBc/Hz number. Corrections for the analyzer's noise power measurement and noise bandwidth must be included, as described in [12]. For example, when noise is measured with the HP3580A spectrum analyzer with 30-Hz resolution bandwidth, the correction is −19.1 dB [13]. The SSB phase noise of the signal generator reference must be much better than the device under test.

    In addition to theoretical concerns, the practical implementation must take the following into account:

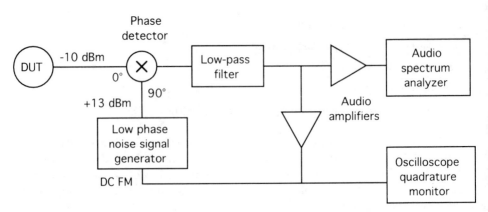

Figure 3.33 Measurement of SSB phase noise by the phase detector method.

- The device under test must be adequately isolated from the setup; otherwise, it will injection-lock to the reference signal leaking back through the mixer-phase detector. The symptoms of this are that the beat note when the test setup is not phase-locked is not sinusoidal, and zero dc cannot be obtained at phase lock. Buffer amplifiers, attenuators, and double-shielded coaxial cables all help isolate the test oscillator from the measurement setup.
- The power supplies feeding the device under test, as well as those supplying the low-noise amplifiers, must be especially free of noise. Power line 60-Hz sidebands, if present, can saturate low-noise amplifiers and spectrum analyzer.
- The dc FM deviation setting of the signal generator affects the loop bandwidth. Measurements of SSB phase noise at offsets less than the deviation setting cannot be performed. Also, inversion of the signal feeding the dc FM signal generator input may required, depending on the sign of the phase detector output.
- A proper calibration method must be devised. Offset the signal generator frequency past the lock range of the dc FM loop (or open the loop), so that a beat note appears on the spectrum analyzer at the offset frequency that the SSB phase noise is to be measured. This becomes the 0-dB reference to which any correction factors having to do with attenuator settings and noise measurement must be applied.
- Measurements down to −160 dBc/Hz with an uncertainty of less than 3 dB are possible with this method.

REFERENCES

[1] ingSOFT Ltd, *RFDesigner™ Software for RF Analysis and Optimization*, Software Manual, "RF Handbook" section. Willowdale, ONT, 1990.

[2] Ansoft Corporation, Four Station Square, Suite 660, Pittsburgh, PA 15219, (412) 261-3200.

[3] ingSOFT Ltd, *RFLaplace*, 213 Dunview Ave., Willowdale, ONT M2N 4H9, 1993, Canada, 416-730-9611.

[4] Wadell, B. C., *Transmission Line Design Handbook*. Norwood, MA: Artech House, 1991.

[5] Goldman, S., "Divider Delay: The Missing PLL Analysis Ingredient," *Frequency Synthesis Handbook*, a collection of articles from *RF Design* magazine. Englewood, CO: Cardiff Publishing Co., 1992, p. 23.

[6] Rohde, U. L., *Digital PLL Frequency Synthesizers — Theory and Design*. Englewood Cliffs, NJ: Prentice-Hall, 1983.

[7] Manassewitsch, V., *Frequency Synthesizers — Theory and Design*. New York: John Wiley and Sons, 1976.

[8] Snow, P., "Use a Spectrum Analyzer to Find Oscillator VSWR," *Microwaves & RF*, April 1985, pp. 109, 114.

[9] Warner, F., and G. Hobson, "Loaded Q Factor Measurements on Gunn Oscillators," *Microwave Journal*, January 1970, pp. 50–53.

[10] Robins, W. P., *Phase Noise in Signal Sources*. London: Peter Peregrinus Ltd, 1982, p. 53.

[11] Rhea, R. W., *Oscillator Design and Computer Simulation*. Englewood Cliffs, NJ: Prentice-Hall, 1990, p. 32.

[12] Hewlett Packard, *Phase Noise Characterization of Microwave Oscillators,* Product Note 11729B-1, 1984.

[13] Thomas, G., *Measurement of Single Sideband Noise.* Motorola Canada Internal Publication, North York, Ontario, November 11, 1983.

CHAPTER 4

Useful Formulas

4.1 ANTENNAS

Directivity (directive gain) is based only on the antenna radiation pattern and does not take into account the antenna efficiency:

$$D = \frac{P_{max}}{P_{av}}$$

D = directivity, linear ratio ≥ 1

P_{max} = maximum power density in any direction (W/m² or W/sr)

P_{av} = average radiated power density (W/m² or W/sr)

Common antenna directivities:

Dipole	1.5 or 1.76 dB
Monopole over ground plane	1.5 or 1.76 dB
Parabolic dish	$D \approx \dfrac{(\pi d)^2}{\lambda^2}$
Horn antenna	$D \approx \dfrac{10A}{\lambda^2}$

D = directive gain (linear)

d = diameter of parabola (m)

λ = wavelength (m)

A = area of horn antenna flange (m²)

Power gain includes the effect of antenna efficiency:

$$G_P = \frac{4\pi P_a}{P} = kD$$

G_P = power gain, linear ratio

P_a = maximum power radiated per unit solid angle

P = power fed into a matched antenna

k = antenna efficiency (see below)

D = antenna directivity

Effective area of antenna:

$$A_e = \frac{G\lambda^2}{4\pi} \tag{4.1}$$

A_e = antenna effective area (m²)

G = gain over isotropic (linear)

λ = wavelength of radiated signal (m)

Antenna efficiency k:

$$k = \frac{\text{Radiated power}}{\text{Input power}} = \frac{R_r}{R_r + R_{\text{loss}}}$$

R_r = radiation resistance (Ω)

R_{loss} = ohmic loss resistance (Ω)

Typical efficiency for a parabolic dish is near 0.5, so that its gain would be about one-half its directivity.

Friis transmission formula:

$$\frac{P_r}{P_t} = \frac{G_r G_t \lambda^2}{(4\pi r)^2} \tag{4.2}$$

P_r = received power (W)

P_t = transmitted power (W)

G_r = power gain of receiving antenna (linear)

G_t = power gain of transmitting antenna (linear)

λ = wavelength (m)

r = distance between receiver and transmitter (m)

Equation (4.2) assumes free-space plane wave transmission and no polarization mismatch between receiving and transmitting antennas.

If we know the antenna effective area and received power, we can obtain the received power density and field strength.

$$P_D = \frac{P_r}{A_e} = \frac{E_r^2}{120\pi} \tag{4.3}$$

$$E_r = \sqrt{P_D 120\pi} \tag{4.4}$$

P_D = received power density (W/m²)

P_r = received power (W)

A_e = receiving antenna effective area (m²)

E_r = received electric field strength, rms (V/m)

The assumptions are still that energy is propagated through a plane wave in free space and that there is no polarization mismatch between the plane wave and the receiving antenna.

The received electric field strength figures prominently in many specifications; we can combine (4.1), (4.3), and (4.4) to come up with the field strength value from a measurement of received power knowing the frequency and the antenna gain:

$$E_r = 2.294 \times 10^{-7} f \sqrt{\frac{P_r}{G_r}} = \sqrt{\frac{30 P_t G_t}{r^2}} \tag{4.5}$$

E_r = received electric field strength, rms (V/m)

f = frequency (Hz)

P_r = received power (W)

G_r = receiver antenna gain over isotropic (linear)

P_t = transmitted power (W)

G_t = transmitter antenna gain over isotropic (linear)

r = distance (m)

4.2 DOPPLER SHIFT

The Doppler effect is a shift in observed frequency when transmitter and receiver are moving relative to each other [1]. Two different types of Doppler shift are of importance: *communication Doppler shift* and *radar Doppler shift*. The observed frequency of an RF source in the presence of relative motion is given by (4.6).

$$f_0 = f_s \frac{1 - \dfrac{v \cos\phi}{c}}{\sqrt{1 - \dfrac{v^2}{c^2}}} \qquad (4.6)$$

f_0 = observed frequency (Hz)

f_s = frequency if RF source were stationary relative to observer (Hz)

v = relative velocity of RF source (m/s)

ϕ = angle between relative velocity and line of sight between source and observer

A more convenient parameter is the Doppler shift, the amount of frequency shift that can be attributed to relative motion:

$$\Delta f \approx f_s \frac{v}{c} \cos(\phi) \qquad (4.7)$$

Δf = frequency shift due to relative motion

f_s = frequency if object were stationary (Hz)

v = relative velocity of object (m/s)

c = speed of light, 2.9979×10^8 m/s

ϕ = angle between relative velocity and line of sight between source and observer (bearing angle)

The frequency increases for approaching objects and decreases for receding objects.

If two mobile radios operating at 900 MHz are receding from each other at 100 km/hr, their relative velocity is 200 km/hr, or 55.6 m/s, resulting in a frequency shift of 167 Hz in the carrier frequency of one relative to the other. This is the communication Doppler shift.

The radar Doppler shift [2,3] will exhibit *double* the frequency shift of the communication shift, because the incident frequency is already shifted by the amount of the communication Doppler shift as it hits the target and is then shifted again upon reflection from the target.

$$\Delta f_{\text{radar}} \approx 2 \frac{f_s v}{c} \cos(\phi) \qquad (4.8)$$

Δf_{radar} = frequency shift of reflected radar signal due to motion of target

$\quad$ f_s = frequency if object were stationary (Hz)

$\quad$ v = relative velocity of object (m/s)

$\quad$ c = speed of light, 2.9979×10^8 m/s

$\quad$ ϕ = angle between relative velocity and line of sight between source and observer (bearing angle)

4.3 ERROR FUNCTION

Error function is used to determine the probability of Gaussian random events. If we have a Gaussian process with a mean value of μ and standard deviation σ, then the probability of an observed value y being greater than $(\mu + x\sigma)$ is $Q(x)$:

$$p[y > (\mu + x\sigma)] = Q(x) = \frac{1}{\sqrt{2\pi}}\int_x^\infty e^{-\lambda^2/2}d\lambda = \frac{1}{2}\,\text{erfc}\left(\frac{x}{\sqrt{2}}\right)$$

For $x > 2$, this function can be approximated by:

$$p[y > (\mu + x\sigma)] \approx \frac{1}{x\sqrt{2\pi}}\left(1 - \frac{1}{x^2}\right)e^{-x^2/2} \qquad x > 2$$

$\qquad$ Definition of the error function has not been entirely consistent in the literature; the definitions differ in scaling and multiplication constants. While most authors dismiss this simply as a difference in approach, numerical calculations require rigorous and consistent definitions. Throughout this book, the following definition of the error function has been used:

$$\text{erf}(x) = \frac{2}{\sqrt{\pi}}\int_{-\infty}^x e^{-\lambda^2}d\lambda$$

$$\text{erfc}(x) = 1 - \text{erf}(x) = \frac{2}{\sqrt{\pi}}\int_x^\infty e^{-\lambda^2}d\lambda$$

4.4 FREQUENCY AND IMPEDANCE SCALING

If we have a passive circuit operating at frequency f_1 and impedance level R_1 and want to transform it to a new, different frequency of operation f_2 and impedance level R_2, the following transformations are applicable:

$$L' = L\frac{R_2}{R_1}\frac{f_1}{f_2} \tag{4.9}$$

$$C' = C\frac{R_1}{R_2}\frac{f_1}{f_2} \tag{4.10}$$

$$R' = R\frac{R_2}{R_1} \tag{4.11}$$

$$Z_0' = Z_0\frac{R_2}{R_1} \tag{4.12}$$

$$\text{length}' = \text{length}\frac{f_1}{f_2} \tag{4.13}$$

L = inductance in original circuit operating at f_1 and R_1

L' = inductance in new, transformed circuit operating at f_2 and R_2

C = capacitance in original circuit operating at f_1 and R_1

C' = capacitance in new, transformed circuit operating at f_2 and R_2

R_1 = impedance level, terminating impedance in original circuit

R_2 = impedance level, terminating impedance in new, transformed circuit

R = resistance in original circuit operating at f_1 and R_1

R' = resistance in new, transformed circuit operating at f_2 and R_2

f_1 = corner, resonant, or reference frequency of original circuit

f_2 = corner, resonant, or reference frequency of new circuit

Z_0 = characteristic impedance in original circuit operating at f_1 and R_1

Z_0' = characteristic impedance in new, transformed circuit operating at f_2 and R_2

length = transmission line length in original circuit operating at f_1 and R_1

length' = transmission line length in new, transformed circuit operating at f_2 and R_2.

Invariant quantities in frequency and impedance scaling are transformer turn ratio, tap location in tapped coils, and TEM transmission line velocity factor.

Figure 4.1 illustrates low-pass to high-pass, band-pass, and band-reject transformations, starting from a low-pass filter normalized to $\omega = 1$, $R = 1$.

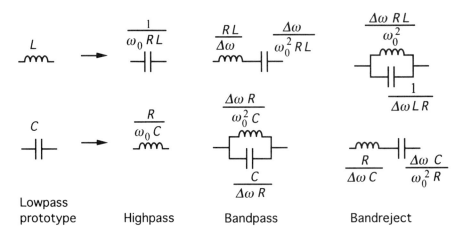

Lowpass prototype   Highpass   Bandpass   Bandreject

Figure 4.1 High-pass, bandpass, and bandreject transformations.

$\Delta\omega$ = 3-dB radian bandwidth = $2\pi BW_{\text{Hz}}$ (rad/s)

ω_0 = radian center frequency for bandpass and bandreject, 3-dB corner frequency for highpass (rad/s)

R = terminating resistance required in transformed filter (Ω)

4.5 GAIN

Gain has many meanings, depending on the context. The most generally useful concept of gain is *transducer power gain,* defined as the ratio between power delivered to the load and power available from the source. The next most useful variation is *insertion gain,* defined as the ratio between power delivered to the load when directly connected to the source and power delivered to the load through the device under test. These two gains are equal to each other if the load and the source impedances are equal and the device under test is conjugately matched at both ports. The same discussion applies to loss.

$$G_T = \frac{4R_S}{R_L}\left|\frac{V_L}{V_S}\right|^2 \tag{4.14}$$

$$G_I = \left(\frac{R_S}{R_L}+1\right)^2\left|\frac{V_L}{V_S}\right|^2$$

G_T = transducer power gain (linear)

G_I = insertion power gain (linear)

R_S = source resistance (Ω)

R_L = load resistance (Ω)

V_L = voltage across load (V)

V_S = source open-circuit voltage (V)

Transducer gain can be obtained from the *s*-parameters, if the source and load reflection coefficients are known.

$$G_T = \frac{|s_{21}|^2(1 - |\Gamma_L|^2)(1 - |\Gamma_S|^2)}{|1 - s_{11}\Gamma_S - s_{22}\Gamma_L + \Delta\Gamma_S\Gamma_L|^2} \qquad (4.15)$$

$$\Delta = s_{11}s_{22} - s_{12}s_{21}$$

G_T = transducer gain, (linear)

Γ_S = source reflection coefficient (complex)

Γ_L = load reflection coefficient (complex)

The following example illustrates the usage of transducer gain in an attenuator application, where the concept of gain or loss between source and load of different impedances can get confusing. Consider the attenuator shown in Figure 4.2, operating from a 50-Ω source into a 75-Ω load.

This attenuator is matched to 50 Ω at its input and 75 Ω at its output; output voltage is half of input voltage. Therefore, it would appear that this is a 6-dB attenuator. However, if we look at power, the power available from the input is $0.5^2/50$ = 5 mW, while the power absorbed by the load is $0.25^2/75$ = 0.8333 mW. The ratio of power absorbed by load to power available from source is 0.1666, which is –7.8 dB and is the correct answer for transducer gain.

RF and microwave circuits are much more concerned with power than voltage, because ultimately it is power that gets radiated out of the antenna or conducted

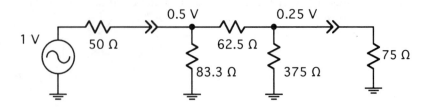

Figure 4.2 A 50-Ω to 75-Ω attenuator.

down the cable, and it is ultimately noise power that limits receiver sensitivity. Therefore, transducer gain is the most commonly used gain concept.

4.6 GROUP DELAY

Group delay is the slope of phase versus frequency. Group delay variations across the modulation bandwidth can introduce distortion into an otherwise perfectly linear system. When modulation sidebands are delayed with respect to each other, the demodulated waveform is distorted.

$$\tau = -\frac{d\varphi\,[\text{rad}]}{d\omega\,[\text{rad/s}]} = -\frac{1}{360}\frac{d\phi\,[\text{deg}]}{df\,[\text{Hz}]} \approx -\frac{1}{360}\frac{\Delta\phi\,[\text{deg}]}{\Delta f\,[\text{Hz}]} \tag{4.16}$$

τ = group delay (s)

φ = angle, phase of measured quantity (rad)

ϕ = angle, phase of measured quantity (degrees)

ω = radian frequency, (rad/s)

f = frequency (Hz)

It can be shown that group delay and insertion loss shapes are related. In a bandpass filter, the most power is absorbed in the filter at the peaks of group delay. This is frequently important in crystal filters, which are very sensitive to power levels. To measure the worst-case power absorbed in a crystal filter, find the frequency of highest group delay, which will be in the filter pass band, and measure the incident, reflected, and transmitted powers at that frequency. By the law of conservation of energy, the power absorbed by the filter (its loss) will be the difference between the incident power and the sum of reflected and transmitted powers.

Group delay through an oscillator resonator network can be related to its loaded Q by (3.27) and thus to the oscillator SSB phase noise.

4.7 IMPEDANCE, REFLECTION COEFFICIENT, VSWR

Impedance and reflection coefficient are related quantities provided the reference impedance Z_0 is known.

$$\Gamma = \frac{Z - Z_0}{Z + Z_0} \tag{4.17}$$

$$Z = \frac{1 + \Gamma}{1 - \Gamma}Z_0 \tag{4.18}$$

Γ = reflection coefficient (complex, dimensionless)

Z = any impedance (complex) (Ω)

Z_0 = reference (characteristic) impedance (real) (Ω)

The reflection coefficient can also be defined for complex reference impedance. Such a definition is useful in evaluating interstage mismatch, where the source
and load impedances are complex. In general,

$$\Gamma = \frac{Z_L - Z_S^*}{Z_L + Z_S} \qquad (4.19)$$

Γ = reflection coefficient (complex, dimensionless)

Z_L = load impedance (complex) (Ω)

Z_S = reference (source) impedance (complex) (Ω)

* = complex conjugate

Equation (4.17) is, then, a special case of (4.19), where $Z_0 = Z_S$ is real and is
therefore its own complex conjugate.

$$\text{Return loss} = -20 \log(|\Gamma|) \qquad (4.20)$$

Γ = reflection coefficient

$|\Gamma|$ = magnitude of reflection coefficient

$$VSWR = \frac{1 + |\Gamma|}{1 - |\Gamma|} = \frac{V_{max}}{V_{min}} \qquad (4.21)$$

$VSWR$ = voltage standing wave ratio (always > 1) (linear)

Γ = voltage reflection coefficient (linear)

$|\Gamma|$ = magnitude of reflection coefficient

V_{max} = rms voltage at a point of maximum voltage (V)

V_{min} = rms voltage at a point of minimum voltage (V)

$VSWR$ causes transmission loss to increase.

$$\alpha_r = \alpha \left[\frac{(VSWR)^2 + 1}{2(VSWR)} \right] \qquad (4.22)$$

α_r = corrected attenuation (Nepers/m)

α = attenuation in a matched line (Nepers/m)

VSWR = measure of mismatch on a transmission line

The general equation for the input impedance of a terminated transmission line:

$$Z = Z_0 \frac{Z_L + Z_0 \tanh(\gamma l)}{Z_0 + Z_L \tanh(\gamma l)} \qquad (4.23)$$

Or, if line is lossless:

$$Z = Z_0 \frac{Z_L + jZ_0 \tan(\beta l)}{Z_0 + jZ_L \tan(\beta l)} \qquad (4.24)$$

Z = input impedance of transmission line (complex) (Ω)

Z_0 = characteristic impedance of transmission line, real (Ω)

Z_L = terminating impedance (complex) (Ω)

tanh = hyperbolic tangent

γ = propagation constant, $\alpha + j\beta$

α = attenuation constant (Nepers/m)

β = phase constant = $\omega/v = \omega/(v_p c) = 2\pi/\lambda = 2\pi f/(v_p c)$

l = length of transmission line, (m)

The hyperbolic tangent of a complex quantity can be rather laboriously simplified by use of the following identity:

$$\tanh(x \pm jy) = \frac{\sinh(2x)}{\cosh(2x) + \cos(2y)} \pm j\frac{\sin(2y)}{\cosh(2x) + \cos(2y)} \qquad (4.25)$$

This form is useful, because it allows the expansion of (4.23) into real and imaginary components, which can then be used in a Q calculation, for example.

4.8 INDUCTORS

4.8.1 Air-Wound Coils

Inductors are fundamental components in RF circuits. Their use at very low frequencies is undesirable due to their large size and high losses; these properties are somewhat less restrictive at frequencies between 10 MHz and 1 GHz, where inductors in the form of air-wound coils are widely used as practical components in filters, impedance-matching circuits, and RF-blocking components. Despite their simple construction, air-wound coils present challenges in high-volume manufacturing: They are difficult to handle with automated vacuum pickup equipment, the leads are not always free of insulation, and they have too much dimensional variability. Some of these problems are avoided by winding the wire on a stable coil form or by encasing it in low-loss plastic material. Desirable air-wound coil properties are high Q, high self-resonant frequency (low parasitic parallel capacitance), and well-controlled inductance value. The usual tolerance on inductance value is 10% to 20%, with 5% tolerance becoming increasingly available at higher cost.

The approximate inductance formula [4,5], which neglects the wire gage and is valid for $A \geq 0.4B$ (long coils; see Figure 4.3.) is

$$L \approx \frac{B^2 n^2}{0.45B + A} \qquad (4.26)$$

L = coil inductance (nH)

B = average coil diameter (mm)

n = number of turns

A = coil length (mm)

All coils have stray parallel capacitance, which limits the frequency up to which the coil behaves as an inductor, and losses that determine the coil Q. The capacitance between turns can be approximated by the capacitance between two parallel wires:

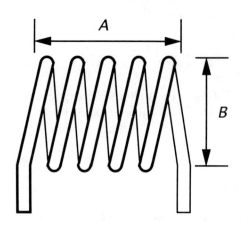

Figure 4.3 Air-wound coil.

$$C \approx \frac{B\epsilon_{\mathit{eff}}}{11.45 \ \cosh^{-1}\left(\dfrac{s}{d}\right)}$$

C = distributed capacitance per turn (pF).

ϵ_{eff} = effective dielectric constant between turns

s = spacing between turns at wire centers (mm)

d = wire diameter (mm), $d < s$

The total capacitance, however, is not $C/(n-1)$, because coupling among nonadjacent turns cannot be neglected. Coupling between first and last turn ($s = A$) sets a lower limit on stray capacitance.

Past its self-resonant frequency, the coil behaves as a capacitor (see Section 3.2 for measurement of coil parallel capacitance). A comprehensive, frequency-dependent inductor model including shield and core losses is discussed in [6]. For most practical purposes, the assumption of constant Q with frequency represents a good compromise between accuracy and complexity.

$$Q = \frac{\omega L}{R_S} \tag{4.27}$$

Q = coil quality factor

ω = radian frequency of measurement ($= 2\pi f$) (rad/s)

f = frequency of measurement (Hz)

L = coil inductance (H)

R_S = series equivalent resistance (Ω)

The highest coil Q is obtained when the ratio between wire diameter and distance between turns is between 0.5 and 0.75, when the length-to-diameter ratio is near 0.96, and when metal surfaces (shields and grounds) are as far away as practical. Typical Qs for air-wound coils are in the range 80 to 200, Q for surface-mount coils wound on ceramic coil forms is from 20 to 60.

4.8.2 Spiral Inductors

Practical air-wound coils can be designed for a Q of up to 150. Such coils need to be space-wound with gaps between turns; this construction results in wide variation (up to ±20%) on their inductance value. Spiral inductors, etched on the circuit board itself, represent another possible compromise, where the Q is sacrificed for excellent inductance value repeatability. Inductors of less than 1% tolerance on their inductance value and Q of 50 are easily achievable.

Both circular and square spiral inductors are feasible. The square inductor in Figure 4.4 has higher inductance for a given circuit board area.

The dc inductance without a ground plane is approximately:

$$L_{DC} \approx 85 \times 10^{-10} Dn^{(5/3)} \qquad (4.28)$$

L_{DC} = low-frequency inductance (H)

D = length of square side (cm)

n = number of turns

$W = S$, strip width and spacing are the same

The inductance with a ground plane on a standard 1.5-mm-thick G-10 circuit board is about 10% to15% lower than the formula in (4.28). The high-frequency equivalent circuit of a printed spiral inductor is shown in Figure 4.5.

Series resistor R and coupling capacitor C_b are frequency dependent. References [7,8] deal with high-frequency properties of spiral inductors in more detail. The main drawback of spiral inductors is their low Q when compared to air-wound coils occupying comparable circuit board area. Their main advantages are near zero cost, low tolerance value, excellent repeatability, good temperature performance, and low susceptibility to mechanical vibration.

4.9 INTERCEPT POINT

Input intercept point is a mathematical tool for evaluating how much distortion will be produced at any input amplitude level, once the distortion at one particular

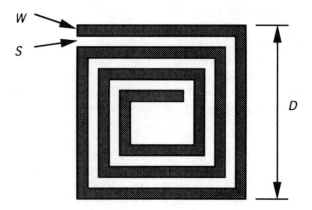

Figure 4.4 Square spiral inductor.

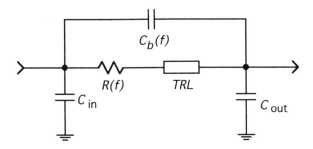

Figure 4.5 High-frequency equivalent circuit of spiral inductor.

input level is known. Similarly, output intercept point is a mathematical tool for evaluating how much distortion will be produced at any output amplitude level, once the distortion at one particular output level is known. The input and output intercept points differ from each other by the system or device gain. The use of one or the other is governed by the actual application: input intercept point is normally used for receiver work, while output intercept point is more useful in cable communication. The input intercept point is used throughout this book for consistency.

IP2 is important in predicting the level of second harmonic generated or for causing a particular receiver spurious response called the *half IF* spurious response. IP3 can be used to predict the level of third harmonic, the amount of IM distortion, and the composite triple beat level. Higher-order intercept points are important for predicting high-order receiver spurious responses.

$$IP_n = A + \frac{\Delta}{n-1} \tag{4.29}$$

IPn = nth-order input intercept point (dBm)

A = input signal level (dBm)

Δ = difference between desired signal and undesired nth-order

distortion (dB)

n = order of distortion

Equation (4.29) is evaluated from measured data. The undesired distortion Δ at some input level A is measured (see Section 3.6). From that information, we can obtain Δ for any other level, provided the order of the distortion product is known. The order n, if not known, can also be obtained by measurement. Simply drop the input level A by 1 dB and observe what happens to the relative difference, Δ. If Δ increases by 2 dB, then $n = 3$ and we are measuring the IP3. The change in Δ following a 1-dB change in input level A is always equal to $(n - 1)$ dB. This means that the distortion products decrease in absolute level by n dB, following a 1-dB decrease in input level.

$$1/2 \text{ IF rejection} = \frac{1}{2}(\text{IP2} - \text{S} - \text{CR}) \qquad (4.30)$$

   IP2 = input second order intercept point (dBm)

     S = receiver sensitivity (dBm)

   CR = capture ratio, or co-channel rejection (dB)

$$\text{IM} = \frac{1}{3}(2\text{IP3} - 2\text{S} - \text{CR}) \qquad (4.31)$$

   IM = intermodulation rejection (dB)

  IP3 = input third order intercept point (dBm)

    S = receiver sensitivity (dBm)

   CR = capture ratio, or co-channel rejection (dB)

4.10 LINE-OF-SIGHT COMMUNICATION

Line-of-sight communication assumes free-space transmission, no polarization mismatch between receiving and transmitting antennas, and no additional loss in the intervening free space.

$$P_r = P_t \frac{G_r G_t \lambda^2}{(4\pi r)^2} \qquad (4.32)$$

  P_r = received power (W)

  P_t = transmitted power (W)

  G_r = power gain of receiving antenna (linear)

  G_t = power gain of transmitting antenna (linear)

  λ = wavelength (m)

   r = distance between receiver and transmitter (m)

    Line-of-sight communication in free space is quite efficient. A 50-W transmitter and a receiver of 0.25-μV sensitivity can communicate over 15,000 km at 300 MHz with unity gain antennas! The reason that terrestrial communications have less range is because propagation along the Earth's surface is limited by curvature and obstacles.

4.11 MISMATCH ERRORS

The majority of formulas mentioned in this book assume that circuit stages are conjugately matched to each other for maximum power transfer. When this condition is not true, as may be the case when examining system performance at the image frequency, for example, the mismatch loss can be taken into account by increasing stage loss (or decreasing stage gain) by the amount of mismatch loss.

We need to define a new reflection coefficient, where the reference impedance is not 50 Ω but the complex source impedance. Figure 4.6 shows the interface between two arbitrary stages. The equations for mismatch loss in decibels:

$$\text{Mismatch loss} = -10 \log(1 - |\Gamma_m|^2) \tag{4.33}$$

$$\text{Mismatch loss} = -10 \log\left[1 - \left(\frac{VSWR - 1}{VSWR + 1}\right)^2\right] \tag{4.34}$$

where

$$\Gamma_m = \frac{Z_L - Z_S^*}{Z_L + Z_S}$$

Γ_m = reflection coefficient (complex, dimensionless)

Z_L = load impedance (complex) (Ω)

Z_S = reference (source) impedance (complex) (Ω)

$VSWR$ = voltage standing wave ratio

* = complex conjugate

By inspection, if Z_S and Z_L are complex conjugates, then the reflection coefficient Γ_m is zero, and mismatch loss is also zero.

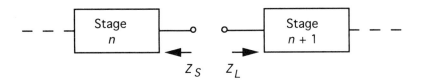

Figure 4.6 Two mismatched stages.

Mismatch loss is the loss between two stages that are not impedance-matched to each other. It is equal to the transmission loss only if all the loss is due to reflection from an impedance mismatch. Device gain must be lowered by the amount of mismatch loss in system calculations, such as receiver sensitivity and intercept point calculations.

4.12 NETWORK TRANSFORMATIONS

4.12.1 Norton Transformation

Norton transformation can be used to decrease the number of components in a filter or a matching network by taking advantage of mutual inductance implemented as a tapped coil. It can also be used to implement or absorb negative component values in special cases. For example, through use of Norton transformation the left-hand circuit in Figure 4.7 can be transformed into the right-hand one [9]. The necessary equations are

$$K = \frac{C_a + C_b}{C_a} = \frac{C_1 + C_2}{C_1}$$

$$C_1 = \frac{C_a}{K}$$

$$C_2 = \frac{C_b}{K}$$

$$tap = \frac{100}{K}$$

To illustrate Norton transformation, we will consider the classical elliptic low-pass topology shown in Figure 4.8.

The filter consists of two coils and five capacitors; usually the capacitor values are widely different in value. Norton transformation can be used to transform the classical topology into one with two coils and only three capacitors, whose values are much closer to each other than the original untransformed values. The transformed filter is shown in Figure 4.9.

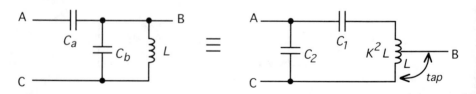

Figure 4.7 Norton transformation.

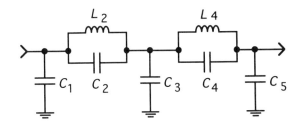

Figure 4.8 Elliptic low–pass prototype filter.

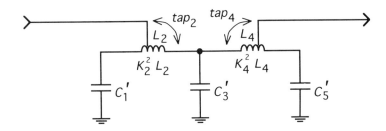

Figure 4.9 Equivalent transformed filter with fewer components.

$$K_2 = \frac{C_2 + C_1}{C_1}$$

$$K_4 = \frac{C_4 + C_5}{C_5}$$

$$C_1' = \frac{C_1}{K_2}$$

$$C_5' = \frac{C_5}{K_4}$$

$$C_3' = C_3 + \frac{C_2}{K_2} + \frac{C_4}{K_4}$$

$$tap_2 = \frac{100}{K_2}$$

$$tap_4 = \frac{100}{K_4}$$

4.12.2 Kuroda Transform

The Kuroda transform (see Figure 4.10) can be used to change a series shorted transmission line stub into a parallel open transmission line stub and vice versa.

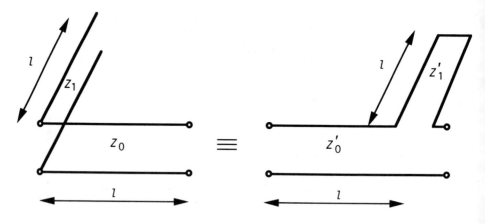

Figure 4.10 Kuroda transformation.

This transformation is commonly used in the design of transmission line filters, where series connections are generally not realizable and shunt connections are.

$$Z_0' = \frac{Z_0}{1 + \dfrac{Z_0}{Z_1}}$$

$$Z_1' = \frac{Z_0^2}{Z_1 + Z_0}$$

$$Z_0 = Z_0' + Z_1'$$

$$Z_1 = \frac{Z_0'(Z_0' + Z_1')}{Z_1'}$$

Z_0, Z_0' = transmission line of length l

Z_1 = shunt open stub of length l

Z_1' = series shorted stub of length l

l = lengths of all transmission lines (all the same length)

4.12.3 Pi to T Transformation

The T-circuit to pi-circuit transformation in Figure 4.11 changes the ordering of series and shunt components in a network, similar to the two previous transformations. Note that the T-network has more circuit nodes than the equivalent pi-network, and thus the transformation forms the basis for many node-reduction algorithms useful in circuit analysis computer programs.

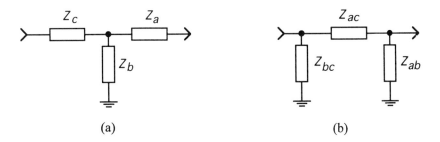

Figure 4.11 Two equivalent networks: (a) T, (b) pi.

$$Z_{ab} = \frac{Z_a Z_b + Z_b Z_c + Z_a Z_c}{Z_c} \tag{4.35}$$

$$Z_{bc} = \frac{Z_a Z_b + Z_b Z_c + Z_a Z_c}{Z_a} \tag{4.36}$$

$$Z_{ac} = \frac{Z_a Z_b + Z_b Z_c + Z_a Z_c}{Z_b} \tag{4.37}$$

$$Z_a = \frac{Z_{ab} Z_{ac}}{Z_{ab} + Z_{bc} + Z_{ac}} \tag{4.38}$$

$$Z_b = \frac{Z_{ab} Z_{bc}}{Z_{ab} + Z_{bc} + Z_{ac}} \tag{4.39}$$

$$Z_c = \frac{Z_{ac} Z_{bc}}{Z_{ab} + Z_{bc} + Z_{ac}} \tag{4.40}$$

4.13 NOISE FORMULAS

The treatment of noise in a device or system can take many forms, and four different methods are used to quantify the amount of noise present: noise figure, noise factor, noise temperature, and ENR (excess noise ratio). Noise figure (in decibels) and noise factor (linear) must not be confused with each other.

Noise power is calculated from noise temperature:

$$P_n = kTB \tag{4.41}$$

P_n = maximum available noise power at conjugate match (W)

k = Boltzmann's constant, 1.38×10^{-23} (J/K)

T = absolute temperature (K)

B = equivalent noise bandwidth (Hz)

The power density at room temperature is approximately −174 dBm/Hz.
Noise factor:

$$F = \frac{\left(\dfrac{S}{N}\right)_{\text{IN}}}{\left(\dfrac{S}{N}\right)_{\text{OUT}}} = \frac{T_e}{T_0} + 1 \tag{4.42}$$

Noise figure:

$$F = 10 \log(F) \tag{4.43}$$

The noise figure (in decibels) of a passive, lossy device is equal to its loss in decibels. Correction for physical temperature different from reference temperature is given by (1.4).
Noise temperature:

$$T_e = (F - 1)\, T_0 \tag{4.44}$$

$$F = 1 + T_e / T_0 \tag{4.45}$$

$(S/N)_{\text{IN}}$ = input signal-to-noise ratio (linear)

$(S/N)_{\text{OUT}}$ = output signal-to-noise ratio (linear)

T_e = equivalent noise temperature (K)

T_0 = reference temperature, 290K

F = noise figure (dB)

F = noise factor (linear)

Cascade noise temperature:

$$T_{eq} = T_1 + \frac{T_2}{G_1} + \frac{T_3}{G_1 G_2} + \ldots + \frac{T_n}{G_1 G_2 \ldots G_{n-1}} \tag{4.46}$$

T_{eq} = equivalent input noise temperature (K)

T_j = stage noise temperature (K)

G_k = stage gain (linear)

Cascade noise factor:

$$F_{eq} = F_1 + \frac{F_2 - 1}{G_1} + \frac{F_3 - 1}{G_1 G_2} + \ldots + \frac{F_n - 1}{G_1 G_2 \ldots G_{n-1}} \tag{4.47}$$

F_{eq} = equivalent input noise factor (linear)

F_j = stage noise factor (linear)

G_k = stage gain (linear)

Input signal level for desired $(S/N)_{out}$:

$$S_{in} = kT_0 BF_{eq}(S/N)_{out}$$
$$S_{in} = k(T_{eq} + T_0)B(S/N)_{out}$$

S_{in} = input signal power (W)

k = Boltzmann's constant, 1.38×10^{-23} (J/K)

T_{eq} = equivalent input noise temperature

T_0 = reference temperature, 290K

B = noise equivalent bandwidth (Hz)

F_{eq} = equivalent input noise factor (linear)

$(S/N)_{out}$ = desired output signal-to-noise ratio (linear)

Noise figure of active two-port device (noise circles are given in Subsection 4.19.3):

$$F = F_{min} + 4\frac{R_n}{Z_0}\frac{|\Gamma_S - \Gamma_{opt}|^2}{|1 + \Gamma_{opt}|^2(1 - |\Gamma_S|^2)} \tag{4.48}$$

F = device noise factor (linear)

F_{min} = minimum noise factor (device property) (linear)

R_n = noise resistance (device property) (Ω). Sometimes R_n is already normalized to (i.e., divided by) Z_0.

Z_0 = system impedance (real) (Ω)

Γ_S = source reflection coefficient

Γ_{opt} = source reflection coefficient for minimum noise factor (device property)

RMS noise voltage across a resistor:

$$V_{nOC} = \sqrt{4RkTB} \tag{4.49}$$

V_{nOC} = open-circuit RMS thermal voltage (V)

R = resistor value (Ω)

k = Boltzmann's constant, 1.38×10^{-23} (J/K)

T = resistor temperature (K)

B = bandwidth (Hz)

The maximum noise voltage available to a load resistor of the same value R in a matched system is $V_{nOC}/2$.
Calibrated noise source formulas:

$$\text{ENR} = 10 \log\left(\frac{T_{ENR} - 290}{290}\right) \tag{4.50}$$

$$T_{ENR} = 290(10^{ENR/10} + 1) \tag{4.51}$$

ENR = excess noise ratio of noise source (dB)

T_{ENR} = equivalent noise temperature of noise source (K)

4.14 OSCILLATOR PHASE NOISE

SSB phase noise places a limit on receiver adjacent channel selectivity, affects the ultimate receiver S/N (usually called Hum & Noise), and influences a Doppler radar's velocity resolution. Single-sideband phase noise at an offset f_m from the carrier is governed by Leeson's equation [10]:

$$\mathscr{L}_{PM} \approx 10 \log\left[\frac{FkT}{A} \frac{1}{8Q_L^2} \left(\frac{f_0}{f_m}\right)^2\right] \tag{4.52}$$

$\mathscr{L}_{PM}$ = single-sideband phase noise density (dBc/Hz)

F = device noise factor at operating power level A (linear)

k = Boltzmann's constant, 1.38×10^{-23} ((J/K))

T = temperature (K)

A = oscillator output power (W)

Q_L = loaded Q (dimensionless)

f_0 = oscillator carrier frequency (Hz)

f_m = frequency offset from carrier (Hz)

Equation (4.52) applies provided f_m is greater than the active device's $1/f$ flicker corner frequency; the active device's noise factor at operating power level is known; active device operation is close to linear; the loaded Q includes the effects of component losses, device loading, and output buffer loading; and only a single resonator is used in the oscillator.

In theory, the noise power density is split equally between AM and PM components [10]; the total noise power density is then twice that given by (4.52). In practice, the amount of AM noise is usually far greater than PM noise at frequency offsets far from carrier, while PM noise dominates close to the carrier.

Leeson's equation provides several basic insights into the characteristics of oscillator SSB phase noise: doubling the loaded Q improves SSB phase noise by 6 dB; doubling the operating frequency results in 6 dB SSB phase noise degradation, all other things being equal. This is basically the same degradation as for frequency multiplication. There is a 6-dB/octave slope with respect to the frequency offset f_m, and operation at high signal level is desirable.

The most troublesome variable in Leeson's equation is the device noise factor, which is difficult to estimate because of nonlinear operation. What is usually done in practice is that the loaded Q is obtained using techniques introduced in Section 3.10, SSB phase noise is measured, and then noise factor is calculated using Leeson's equation. It is usually 2 to 3 dB worse than small-signal noise figure listed in the device data sheet.

Leeson's equation only applies between the $1/f$ flicker noise transition frequency (f_1) and a frequency past which amplified white noise dominates (f_2) in Figure 4.12. Typically, f_1 is less than 1 kHz and should be as low as possible; f_2 is

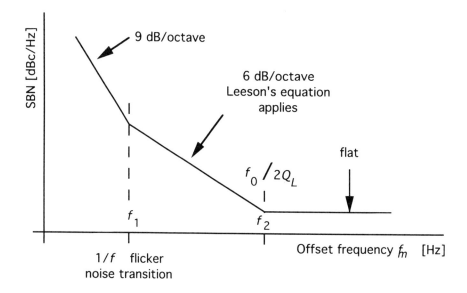

Figure 4.12 Typical oscillator phase noise density as a function of offset frequency.

in the region of a few MHz. High-performance oscillators require devices specially selected for low $1/f$ transition frequency. JFETs have the lowest $1/f$ flicker noise transition frequency, next come bipolar transistors; GaAs FETs are the worst. There is an inverse correlation between dc β and $1/f$ noise of bipolar transistors [11] (higher dc β = lower $1/f$ noise).

The relationship between the unloaded resonator Q_U and loaded oscillator Q_L is important in determining SSB phase noise. In theory, they can be made equal, if lossless components are used, loading due to active device can be neglected, and output power is negligible compared to operating power. In practice, none of these restrictions can be satisfied; $Q_L = 3/4\ Q_U$ is usually the upper limit for a fairly sophisticated oscillator design. In wide-tuning VCOs, where varactor Q contributes to overall Q degradation, the ratio between unloaded and loaded Q can be as high as 4. Loaded Q can also be obtained from computer simulation as the phase slope (i.e., group delay) of the open-loop oscillator equivalent circuit [12], as mentioned in Section 3.10.

Wide-tuning VCOs also suffer from spurious FM modulation of the VCO due to thermal noise in the varactor diode(s). The varactor diode series resistance can be transformed to its parallel equivalent by use of (2.51), which serves as a source of thermal voltage modulating the VCO. This noise voltage is usually filtered by a separate capacitor in the steering line circuit, mounted as close to the VCO as possible.

$$\mathscr{L}_v = 10 \log\left[\frac{2K_v^2 kTR_p}{f_m^2 + (2\pi f_m^2 C_{SL}R_p)^2}\right] \tag{4.53}$$

$\mathscr{L}_v$ = single-sideband noise due to varactor resistance (dBc/Hz)

K_v = steering line sensitivity (Hz/V)

k = Boltzmann's constant, 1.38×10^{-23} (J/K)

T = absolute temperature (K)

R_p = varactor diode equivalent parallel resistance, or total equivalent parallel
   resistance in the steering line circuit (Ω)

f_m = frequency offset from carrier (Hz)

C_{SL} = steering line capacitance to ground (F)

Given a certain varactor parallel resistance R_p and VCO steering line sensitivity K_v, (4.53) places a lower limit on C_{SL} for a certain SSB phase noise goal. C_{SL} must be a low impedance capacitor to ground, with no resistance in series with it. C_{SL} is not the varactor capacitance but the value of an additional steering line capacitor required to minimize SSB phase noise degradation due to steering line resistance. SSB noise of (4.53) is additive to (4.52) on a power basis.

Sideband noise of an operating oscillator can be theoretically improved by the arrangement [13] shown in Figure 4.13.

The long delay τ serves to uncorrelate the noise sidebands in the two signal paths to the mixer, so that upon mixing, the SSB phase noise increases only by 3 dB, rather than the 6 dB incurred by direct frequency multiplication. As division by 2 improves the SSB phase noise by 6 dB, there is a net theoretical improvement in SSB phase noise of 3 dB for frequency offsets greater than $1/\tau$.

4.15 QUALITY FACTOR

The most basic definition of quality factor relates energy stored to the average power loss:

$$Q = \omega_0 \frac{\text{energy stored}}{\text{average power loss}} \tag{4.54}$$

Q = quality factor

ω_0 = resonant frequency (rad/s)

If the network or structure can be represented as a series connection of resistances R_j and reactances X_i, then the Q of such a network can be obtained from (4.55). This expression makes no assumption about the variation of X_i with frequency and is therefore suitable for transmission line resonators, LC circuits, and nonresonant circuits.

$$Q = \frac{1}{2} \frac{\sum_{i=1}^{N} X_i + \omega \sum_{i=1}^{N} \left| \frac{dX_i}{d\omega} \right|}{\sum_{j=1}^{M} R_j} \tag{4.55}$$

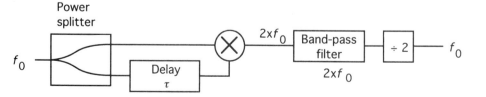

Figure 4.13 Improvement of SSB phase noise by uncorrelated mixing and frequency division.

Q = quality factor of network

i = index of summation for the N series reactances

N = number of reactances

X_i = series-connected reactances

ω = radian frequency (rad/s)

j = index of summation for the M series resistances

M = number of resistances

R_j = series-connected resistances

$\dfrac{dX_i}{d\omega}$ = reactance slope parameter

It can be easily verified that the relationship of (4.55) reduces to (4.56) for inductors and capacitors.

$$\text{In series } LC \text{ resonant circuits: } Q = \frac{X}{R_S} \tag{4.56}$$

$$\text{In parallel } LC \text{ resonant circuits: } Q = \frac{R_P}{X} \tag{4.57}$$

Q = quality factor

X = reactance of either coil or capacitor at resonance (Ω)

R_S = series equivalent resistance in series resonant circuit (Ω)

R_P = parallel equivalent resistance in parallel resonant circuit (Ω)

We can demonstrate the application of (4.55) by calculating the Q of an arbitrary length of shorted transmission line section, which is not resonant at the frequency of interest. The input impedance of a shorted transmission line section is given by

$$Z = Z_0 \tanh(\gamma l) = R + jX$$

where, by use of (4.25),

$$R = \frac{Z_0 \sinh(2\alpha l)}{\cosh(2\alpha l) + \cos(2\beta l)} \tag{4.58}$$

$$X \approx Z_0 \tan(\beta l) \tag{4.59}$$

$$\frac{dX}{d\omega} = \frac{Z_0 \frac{l}{v}}{\cos^2(\beta l)} \tag{4.60}$$

It must be emphasized that Q of a transmission line section is not simply equal to X/R, as it would be for inductors and capacitors, but (4.55) must be used to derive the Q in this case. While it may appear that Q is independent of Z_0, this is not really so, because the line attenuation α is a strong function of Z_0.

$$Q = \frac{1}{2} \frac{X + \omega \frac{dX}{d\omega}}{R} = \frac{1}{2} \frac{\tan(\beta l) + \frac{\beta l}{\cos^2(\beta l)}}{\frac{\sinh(2\alpha l)}{\cosh(2\alpha l) + \cos(2\beta l)}} \tag{4.61}$$

Q = quality factor of a shorted nonresonant transmission line section

ω = radian frequency, = $2\pi f$ (rad/s)

α = line attenuation (Nepers/m)

$\beta = \omega/v$

l = line length, $l \neq \lambda/4$ (m)

v = propagation velocity = $v_p c$ (m/s)

Z_0 = transmission line characteristic impedance (Ω)

Another equation for Q, often used in oscillator work, uses group delay through a resonant circuit to define oscillator loaded Q:

$$Q_L = \omega \frac{\tau}{2} = -\frac{\pi f_0}{360} \frac{d\phi}{df} \tag{4.62}$$

Q_L = oscillator loaded quality factor

ω = radian frequency, = $2\pi f_0$ (rad/s)

τ = group delay (s)

f_0 = frequency at which phase equals zero (Hz)

ϕ = phase of open-loop voltage gain transfer function (degrees)

The term *unloaded Q* is used in cases where a resonant circuit is not loaded by any external terminating impedances. In that case, the Q is determined only by resonator losses.

Loaded Q is used in filter and oscillator work to mean the width of the resonance curve, or phase slope, including the effects of external components. In that case, the Q is determined mostly by the external components. The ratio between unloaded and loaded Qs in a filter can be used to estimate insertion loss. The loaded Q of oscillator circuits determines their SSB phase noise via Leeson's equation (4.52).

Calculations of Q are important for $\lambda/4$ resonators. Figure 4.14 shows how a lossy $\lambda/4$ resonator can be modeled by a lossless line with a parallel resistor.

$$R = \frac{Z_0 \sinh(2\alpha_N l)}{\cosh(2\alpha_N l) - 1} = \frac{Z_0 \sinh\left(\dfrac{\alpha_{dB} l}{4.343}\right)}{\cosh\left(\dfrac{\alpha_{dB} l}{4.343}\right) - 1} \qquad (4.63)$$

R = equivalent parallel resistance in lossless line equivalent circuit (Ω)

Z_0 = characteristic impedance of $\lambda/4$ line (Ω)

α_N = attenuation of line (Nepers/m; to convert from dB/m to Nepers/m, divide by 8.686)

α_{dB} = attenuation of line (dB/m)

l = length of $\lambda/4$ line (m)

We can also derive the unloaded Q of a $\lambda/4$ resonator from the line loss.

$$Q = \frac{\pi f_0}{v_p c \alpha_N} = \frac{\pi}{4 \alpha_N l} = \frac{2.171 \pi}{\alpha_{dB} l} \qquad (4.64)$$

Q = unloaded Q of $\lambda/4$ resonator

f_0 = resonant frequency (Hz)

v_p = velocity factor of line

c = speed of light in free space, 2.9979×10^8 (m/s)

α_N = attenuation of line (Nepers/m; to convert from dB/m to Nepers/m, divide by 8.686)

α_{dB} = attenuation of line (dB/m)

l = length of $\lambda/4$ line (m)

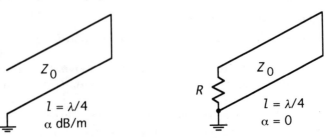

Figure 4.14 Equivalence between lossy line and lossless line with resistor.

The relationship of (4.64) can be inverted to conveniently obtain the line loss from a measurement of $\lambda/4$ resonator Q.

Quality factor of single-pole networks can also be defined and conveniently measured by the width of the resonance curve shown in Figure 4.15:

$$Q = \frac{f_0}{BW_{3dB}} \tag{4.65}$$

f_0 = resonant frequency

BW_{3dB} = 3-dB bandwidth of resonance (same units as f_0)

4.16 RESONATORS

While an inductor and a capacitor can be connected to form a resonant circuit, the term resonator in this section is used to refer to a length of transmission line. The main advantage in using transmission line rather than LC resonators is the much higher Q achievable with transmission line resonators. The high Q can be put to good use in highly selective, low-loss filters, low-noise oscillator circuits, low-loss baluns, and matching networks. Transmission line resonators can usually be made tunable over a wide frequency range with less Q variation than would be present in a tunable inductor.

4.16.1 Coaxial Resonators

A quarter-wave coaxial resonator, shown in Figure 4.16, is formed by shorting the center conductor of a coaxial line to its shield at one end, leaving the other end open-circuited.

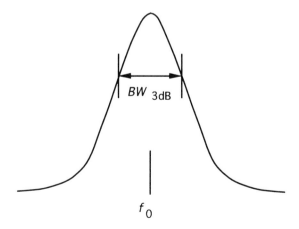

Figure 4.15 Measurement of Q from frequency response.

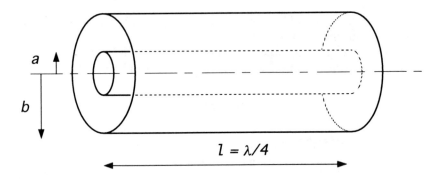

Figure 4.16 Quarter–wave coaxial resonator.

The physical length of a coaxial resonator is equal to one-quarter the wavelength in the medium filling the resonator; the electrical length of a $\lambda/4$ resonator is always 90°. The unloaded quality factor has two contributors: conductor loss, Q_c, and dielectric loss, Q_d. Equation (4.66) gives the general expression for conductor contribution to Q, which is independent of the dielectric, and (4.67) is a simplified expression for the case of copper conductors, with the optimum b/a ratio of 3.59. This optimum ratio yields a 76.7-Ω characteristic impedance in free air; the optimum ratio is also independent of the dielectric constant filling the resonator.

$$Q_c = 2\frac{\sqrt{\pi f \mu \sigma}\ln\left(\frac{b}{a}\right)}{\left(\frac{1}{a}+\frac{1}{b}\right)} \tag{4.66}$$

$$Q_{cc} = 8.398b\sqrt{f} \tag{4.67}$$

$$Q_d = \frac{1}{\tan(\delta)}$$

$$\frac{1}{Q} = \frac{1}{Q_c} + \frac{1}{Q_d} \tag{4.68}$$

$$Z_0 = \frac{60}{\sqrt{\epsilon_r}}\ln\left(\frac{b}{a}\right) \tag{4.69}$$

$$Z_{\text{opt}} = 76.7\sqrt{\epsilon_r} \tag{4.70}$$

$$V_s = \frac{4Q}{n\pi}$$

b = inside radius of outer shield (m)

a = outer radius of inner conductor (m)

f = resonant frequency (Hz)

μ = permeability of conductor (H/m)

σ = conductivity of conductor (mho/m)

ϵ_r = dielectric constant of inside dielectric

Q_c = conductor contribution to unloaded Q

Q_{cc} = conductor contribution to unloaded Q in a copper $\lambda/4$ resonator with the optimum b/a ratio of 3.59112

Q_d = dielectric contribution to unloaded Q

Q = unloaded quality factor of $\lambda/4$ resonator, including conductor and dielectric losses

$\tan(\delta)$ = loss tangent of dielectric material filling resonator

Z_0 = characteristic impedance of coaxial structure (Ω)

Z_{opt} = characteristic impedance that yields highest Q (b/a = 3.59112)

V_s = voltage step-up ratio to top of resonator

n = resonator mode, number of quarter-wavelengths at resonance

Observing that $Q \propto \sqrt{\sigma}$, we can obtain the Q for any other conductor material, once the Q for copper resonator of the same dimensions is determined.

4.16.2 Helical Resonators

Helical resonators find use in the frequency gap between discrete LC resonators and reasonable-size coaxial resonators. The resonator itself is a length of helical transmission line, short-circuited at one end. The unloaded Q is strongly correlated to wire gage. Figure 4.17 shows the proportions of helical resonators with shields of square and circular cross-sections, which will yield the highest Q for the given shield size, using copper as the material of the helix and shield. The relevant equations were first derived by Zverev [14]. The general design is determined by the operating frequency and available physical space. Therefore, the known quantities usually are shield size and resonant frequency. From the shield size, we can derive the helix mean diameter, d, which can then be used in the following equations to arrive at other design parameters.

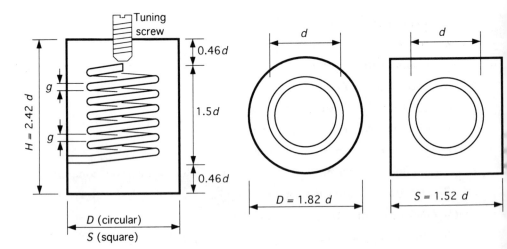

Figure 4.17 Helical resonator dimensions.

$$Q = 35.9 d\sqrt{f}$$

$$N = \frac{2674}{df}$$

$$p = \frac{1759}{d^2 f}$$

$$Z_0 = \frac{136190}{df}$$

$$g = \frac{1}{2p}$$

(4.71)

When the resonator is filled entirely with dielectric, the following parameters change:

$$Z_d = \frac{Z_0}{\sqrt{\epsilon_r}}$$

$$f_d = \frac{f}{\sqrt{\epsilon_r}}$$

$$Q_d = \frac{1}{\dfrac{1}{Q} + \tan(\delta)}$$

(4.72)

d = mean helix diameter (cm)

D = inside diameter of circular shield (cm)

S = length of square shield side (cm)

H = height of shield (cm)

f = resonant frequency (MHz)

f_d = resonant frequency in the presence of dielectric (MHz)

Q = unloaded Q of air-filled resonator

Q_d = unloaded Q in the presence of dielectric material

N = number of turns on helix

p = helix pitch (turns/cm)

Z_0 = characteristic impedance of air-filled helical transmission line (Ω)

Z_d = characteristic impedance with dielectric (Ω)

g = wire diameter; also space between turns (cm)

$\tan(\delta)$ = loss tangent of dielectric material

The input and output coupling method (usually a tap) tends to decrease the resonant frequency of the first and last cavities; shields for those can be made slightly oversize so that the tuning screws come out at an even height for all the resonators in a filter. References [14,15] contain more detailed information on temperature compensation, field analysis, and construction methods. Observing that $Q \propto \sqrt{\sigma}$, we can obtain the Q for any other conductor material, once the Q for copper resonator is determined.

4.16.3 Other Resonators

The resonant frequencies of a metal box, as shown in Figure 4.18, are:

$$f_R = \frac{1}{2\sqrt{\mu\epsilon}}\sqrt{\left(\frac{m}{x}\right)^2 + \left(\frac{n}{y}\right)^2 + \left(\frac{p}{z}\right)^2}$$

f_R = resonant frequency (Hz)

μ = permeability (H/m)

ϵ = permittivity (F/m)

m = integer mode number

n = integer mode number

p = integer mode number

x = x-dimension of box (m)

y = y-dimension of box (m)

z = z-dimension of box (m)

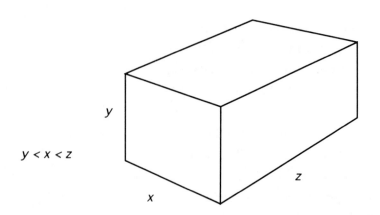

Figure 4.18 Metal box as a waveguide resonator.

The lowest such frequency is for the TE_{101} mode, where $m = 1$, $n = 0$, $p = 1$. Keep in mind that any conductive structure placed inside the box, such as printed circuit board or internal cable, will lower the resonant frequency, analogous to ridged waveguide having lower cutoff frequency than a rectangular waveguide of similar outside dimensions

4.17 SKIN DEPTH

Skin depth is a mathematical simplification that assumes uniform current flow in a certain thickness of conductor δ rather than exponentially decaying current flow, which is the true picture. Nevertheless, this simplification is convenient in estimating the effect of plating thickness on transmission line loss and resonator Q.

$$\delta = \frac{1}{\sqrt{\sigma\pi\mu f}} \tag{4.73}$$

δ = skin depth, approximate depth of penetration of RF current into conductor (m)

σ = conductivity (S/m); see Section 4.24 for common values

μ = permeability (H/m)

f = frequency (Hz)

Several references place an extra factor of $\sqrt{2}$ in the denominator [16,17]. The correctness of our formula can be derived from Maxwell's equations for a plane wave parallel to the surface of a conductor and assuming harmonic variation of E_y with time [18]:

$$\frac{\partial^2 E_y}{\partial x^2} - \gamma^2 E_y = 0$$

where

$$\gamma^2 = j\omega\mu\sigma - \omega^2\mu\epsilon \approx j\omega\mu\sigma$$

because for conductors $\sigma \gg \omega\epsilon$.
The classical solution to the above differential equation is

$$E_y = E_0 e^{-\gamma x}$$

where

$$\gamma = \sqrt{j\omega\mu\sigma} = \sqrt{j}\sqrt{2\pi f\mu\sigma} = \frac{1+j}{\sqrt{2}}\sqrt{2\pi f\mu\sigma} = (1+j)\sqrt{\pi f\mu\sigma}$$

so that

$$E_y = E_0 e^{-\sqrt{\pi f\mu\sigma}x} e^{-j\sqrt{\pi f\mu\sigma}x}$$

and the attenuation factor is

$$e^{-\sqrt{\pi f\mu\sigma}x}$$

Attenuation reaches a value of $e^{-1}(1/e)$ at

$$x = \frac{1}{\sqrt{\pi f\mu\sigma}},$$

which is the definition of skin depth. The discrepancy is due to the evaluation of $\sqrt{j}$.

4.18 SMITH CHART GRAPHICS

With high-end drawing CAD software, such as MiniCAD [19], you can draw your own version of the Smith Chart, or a portion of it, using the equations in this section. The notation has been adopted from [20]. The chart consists of normalized constant r, x, g, and b circles, where $r = 1$, $x = 0$ (as well as $g = 1$, $b = 0$) is the center of the Smith Chart. The coordinate system is rectangular, with u as the x-coordinate (to avoid confusion with x = constant reactance) and v as the y-coordinate such that $(u, v) = (0, 0)$ is the center of the Smith Chart whose radius is 1. Thus, u and v are the equivalent rectangular coordinates of the polar reflection coefficient Γ. Variables used in the following equations are defined at the end of this section.

Constant r circles:

$$\left(u - \frac{r}{1 + r}\right)^2 + v^2 = \frac{1}{(1 + r)^2}$$

This equation defines a family of circles with centers on the u-axis at

$$(u,v) = \left[\frac{r}{(1 + r)}, 0\right]$$

and

$$\text{radius} = \frac{1}{1 + r}$$

Note that $r = 0$ is the unit circle, $u^2 + v^2 = 1$, which is the conventional boundary between inside and outside the Smith Chart. Negative values of r result in circles outside the Smith Chart; $r = -1$ is a circle with infinite radius and is a vertical line at $u = 1$.

Constant x circles:

$$(u - 1)^2 + \left(v - \frac{1}{x}\right)^2 = \frac{1}{x^2}$$

This equation represents circles whose centers are at:

$$(u,v) = \left[1, \pm\frac{1}{x}\right]$$

of

$$\text{radius} = \frac{1}{|x|}$$

Portions of these circles that are inside the Smith Chart are arcs that start at $\alpha_1 = 270°$, continue counterclockwise for a length of α_2 when x is positive:

$$\alpha_2 = -\left[180° - \tan^{-1}\left(\frac{2|x|}{x^2 - 1}\right)\right]$$

When x is negative, the arcs start at $\alpha_1 = 90°$, continue counterclockwise for a length of α_2. (α_2 is thus relative to α_1, not to the center of the Smith Chart):

$$\alpha_2 = 180° - \tan^{-1}\left(\frac{2|x|}{x^2 - 1}\right)$$

Constant g circles:

$$\left(u + \frac{g}{1 + g}\right)^2 + v^2 = \frac{1}{(1 + g)^2}$$

These are circles with centers on the negative u axis at

$$(u,v) = \left[-\frac{g}{(1 + g)}, 0\right]$$

and radius $= 1/(1 + g)$

Constant b circles:

$$(u + 1)^2 + \left(v - \frac{1}{b}\right)^2 = \frac{1}{b^2}$$

This equation represents circles whose centers are at

$$(u,v) = \left[-1, \pm\frac{1}{b}\right]$$

with radius $= 1/|b|$

Portions of these circles that are inside the Smith Chart are arcs that start at $\alpha_1 = 270°$, continue counterclockwise for a length of α_2 when b is positive:

$$\alpha_2 = 180° - \tan^{-1}\left(\frac{2|b|}{b^2 - 1}\right)$$

When b is negative, the arcs start at $\alpha_1 = 90°$, continue counterclockwise for a length of α_2:

$$\alpha_2 = -\left[180° - \tan^{-1}\left(\frac{2|b|}{b^2 - 1} \right) \right]$$

Constant Q circles

$$u^2 + \left(v \pm \frac{1}{Q} \right)^2 = 1 + \frac{1}{Q^2}$$

have centers at

$$(u,v) = \left[0, \pm\frac{1}{Q} \right]$$

$$\text{radius} = \sqrt{1 + \frac{1}{Q^2}}$$

The defining arcs start at α_1 and are of length α_2.

$\alpha_1 = 90° - \tan^{-1}(Q)$ for centers on negative v axis
$\alpha_1 = 270° - \tan^{-1}(Q)$ for centers on positive v axis

In both cases, the extent of the arc is α_2.

$$\alpha_2 = 2 \tan^{-1}(Q)$$

r = normalized resistance

x = normalized reactance

g = normalized conductance

b = normalized susceptance

u = Cartesian (rectangular) x-coordinate

v = Cartesian (rectangular) y-coordinate

α_1 = starting angle of arc, $0° < \alpha_1 < 360°$

α_2 = counterclockwise extent or length of arc, $0° < \alpha_2 < 180°$

$Q = x/r$ quality factor of a given impedance point

Many other properties, in addition to impedances, can be plotted on the Smith Chart. Constant gain, noise figure, and stability circles are defined in the following sections.

4.19 s-PARAMETER FORMULAS

The popularity of s-parameters stems from the fact than open and short circuits are not required in their definition or measurement, as is the case for the corresponding y, z, or h parameters. Well-behaved open and short circuits are difficult to obtain at high frequencies, and measured device stability may be compromised by open- or short- circuit terminations at its terminals. s-parameters are strictly small-signal parameters and are not suitable for describing nonlinear operation. s_{ii} is the reflection coefficient at port i when all other ports are terminated in the characteristic impedance. s_{ij} is the transducer gain from port j to port i under the same conditions.

4.19.1 s-Parameter Formulas: Gain

From the many definitions of gain, the most important one is transducer gain, defined as the ratio between power delivered to the load and power available from the source:

$$G_T = \frac{|s_{21}|^2(1 - |\Gamma_L|^2)(1 - |\Gamma_S|^2)}{|1 - s_{11}\Gamma_S - s_{22}\Gamma_L + \Delta\Gamma_S\Gamma_L|^2} \tag{4.74}$$

$$\Delta = s_{11}s_{22} - s_{12}s_{21}$$

G_T = transducer gain (linear)

Γ_S = source reflection coefficient (complex)

Γ_L = load reflection coefficient (complex)

The maximum of transducer gain, when both source and load are conjugately matched is the maximum available gain:

$$G_{\max} = \left|\frac{s_{21}}{s_{12}}\right|(k - \sqrt{k^2 - 1})$$

$G_{\max}$ = maximum available gain (linear)

k = Rollett's stability factor (see Subsection 4.19.5), $k \geq 1$.

For given values of s_{21} and s_{12}, the highest gain can be obtained for $k = 1$, at the edge of instability. A high k value will limit the maximum available gain. Maximum available gain is difficult to measure directly and is usually computed from the device's s-parameters.

The operating power gain, defined as the ratio between power delivered to the load and power absorbed from the source, is useful for plotting constant gain circles in the output reflection coefficient plane or for analyzing gain available from an unstable device.

$$G_p = \frac{|s_{21}|^2(1 - |\Gamma_L|^2)}{|1 - s_{22}\Gamma_L|^2 - |s_{11} - \Delta\Gamma_L|^2}$$ (4.75)

$$\Delta = s_{11}s_{22} - s_{12}s_{21}$$

G_P = operating power gain (linear)

Γ_L = load reflection coefficient (complex)

Note that the operating power gain is independent of the source reflection coefficient. We can analyze the effect of output termination on operating power gain by plotting constant gain circles on the Smith Chart.

$$\text{Center} = \frac{\frac{G_P}{|s_{21}|^2}(s_{22} - \Delta s_{11}^*)^*}{1 + \frac{G_P}{|s_{21}|^2}(|s_{22}|^2 - |\Delta|^2)}$$ (4.76)

$$\text{Radius} = \frac{\sqrt{\left(\frac{G_P}{|s_{21}|^2}\right)^2|s_{12}s_{21}|^2 - 2k\frac{G_P}{|s_{21}|^2}|s_{12}s_{21}| + 1}}{1 + \frac{G_P}{|s_{21}|^2}(|s_{22}|^2 - |\Delta|^2)}$$ (4.77)

$$\Delta = s_{11}s_{22} - s_{12}s_{21}$$

G_P = power gain represented by circle (linear)

k = Rollett's stability factor (see Subsection 4.19.5)

* = complex conjugate

To maximize transducer gain at the selected operating point, the source reflection coefficient must be the conjugate of the device input reflection coefficient with the selected load connected.

$$\Gamma_S = \left(s_{11} + \frac{s_{12}s_{21}\Gamma_L}{1 - s_{22}\Gamma_L}\right)^*$$

Γ_S = source reflection coefficient for maximum transducer gain

Γ_L = selected load reflection coefficient for desired operating power gain

* = complex conjugate

Constant operating power gain circles are plotted in the output reflection coefficient plane and must not be confused with gain circles, which are sometimes plotted in the input reflection coefficient plane together with noise figure circles. Such gain circles are called *associated gain circles* and refer to gain obtainable when the input is mismatched to obtain specific noise figure.

4.19.2 s-Parameter Formulas: Impedance

Input reflection coefficient for any load impedance:

$$\Gamma_{IN} = s_{11} + \frac{s_{12}s_{21}}{\dfrac{1}{\Gamma_L} - s_{22}} = s_{11} + \frac{s_{12}s_{21}}{\left(\dfrac{Z_L + Z_0}{Z_L - Z_0}\right) - s_{22}} \tag{4.78}$$

Γ_{IN} = reflection coefficient looking into input of device, when output is terminated in Γ_L or Z_L

Γ_L = load reflection coefficient (complex)

Z_L = load impedance (complex) (Ω)

Z_0 = system, reference impedance (real) (Ω)

Output reflection coefficient for any source impedance:

$$\Gamma_{OUT} = s_{22} + \frac{s_{12}s_{21}}{\dfrac{1}{\Gamma_S} - s_{11}} = s_{22} + \frac{s_{12}s_{21}}{\left(\dfrac{Z_S + Z_0}{Z_S - Z_0}\right) - s_{11}} \tag{4.79}$$

Γ_{OUT} = reflection coefficient looking into output of device, when input is terminated in Γ_S or Z_S.

Γ_S = source reflection coefficient (complex)

Z_S = source impedance (complex) (Ω)

Z_0 = system, reference impedance (real) (Ω)

Section 4.19.4 supplies formulas for simultaneous conjugate impedance match at both ports of a device.

4.19.3 s-Parameter Formulas: Noise Circles

$$\text{Center} = \frac{\Gamma_{opt}}{1 + N_i} \tag{4.80}$$

$$\text{Radius} = \frac{1}{1 + N_i}\sqrt{N_i^2 + N_i(1 - |\Gamma_{opt}|^2)} \tag{4.81}$$

where

$$N_i = \frac{(F - F_{min})\,|1 + \Gamma_{opt}|^2}{4\dfrac{R_n}{Z_0}}$$

Γ_{opt} = source reflection coefficient for minimum noise figure (device property) (linear)

F = desired noise factor circle (linear)

F_{min} = minimum noise factor (device property) (linear)

R_n = noise resistance (device property)

Z_0 = reference, system impedance, real (Ω)

N_i = intermediate parameter in calculation

The notation in (4.80) and (4.81) is that of reference [21]. A point on a given noise circle represents the source reflection coefficient that the device needs to see at its input for that noise figure. The device output is conjugately matched for maximum transducer gain:

$$\Gamma_L = \left(s_{22} + \frac{s_{12}s_{21}\Gamma_S}{1 - s_{11}\Gamma_S} \right)^*$$

Γ_S = selected source reflection coefficient for required noise figure

Γ_L = load reflection coefficient for maximum transducer gain

$*$ = complex conjugate

4.19.4 s-Parameter Formulas: Simultaneous Conjugate Match

When $|s_{12}| \neq 0$, the input and output terminations affect the output and input impedances, respectively, and the conditions for simultaneous conjugate match at input and output are not simply s_{11}^* and s_{22}^* but are governed by the following relationships. Simultaneous conjugate match at input and output is required for maximum gain.

$$\Gamma_S = \frac{C_1^*[B_1 - \mathrm{sgn}(B_1)\sqrt{(B_1^2 - 4|C_1|^2)}]}{2|C_1|^2} \tag{4.82}$$

$$\Gamma_L = \frac{C_2^*[B_2 - \mathrm{sgn}(B_2)\sqrt{(B_2^2 - 4|C_2|^2)}]}{2|C_2|^2} \tag{4.83}$$

$$B_1 = 1 + |s_{11}|^2 - |s_{22}|^2 - |\Delta|^2$$
$$B_2 = 1 - |s_{11}|^2 + |s_{22}|^2 - |\Delta|^2$$
$$C_1 = s_{11} - \Delta s_{22}^*$$
$$C_2 = s_{22} - \Delta s_{11}^*$$
$$\Delta = s_{11}s_{22} - s_{12}s_{21}$$

Γ_S = source reflection coefficient for maximum gain

Γ_L = load reflection coefficient for maximum gain

$\text{sgn}()$ = sign function. For stable designs, B_j is positive, and the sign of the square root is therefore negative

$*$ = complex conjugate

4.19.5 *s*-Parameter Formulas: Stability

Stability factor (Rollett's):

$$k = \frac{1 + |s_{11}s_{22} - s_{12}s_{21}|^2 - |s_{11}|^2 - |s_{22}|^2}{2|s_{12}s_{21}|} \tag{4.84}$$

$k > 1$ for unconditional stability.

The Linvill stability factor, C, is the reciprocal of k.

Additional requirements for absolute stability, at which no passive source or load can cause instability:

$$|s_{11}| < 1$$
$$|s_{22}| < 1$$
$$\left| \frac{|s_{12}s_{21}| - |(s_{11} - \Delta s_{22}^*)^*|}{|s_{11}|^2 - |\Delta|^2} \right| > 1$$
$$\left| \frac{|s_{12}s_{21}| - |(s_{22} - \Delta s_{11}^*)^*|}{|s_{22}|^2 - |\Delta|^2} \right| > 1$$
$$\Delta = s_{11}s_{22} - s_{12}s_{21}$$

Input stability circle:

$$\text{Center} = \frac{(s_{11} - \Delta s_{22}^*)^*}{|s_{11}|^2 - |\Delta|^2} \qquad \text{Radius} = \frac{|s_{12}s_{21}|}{|s_{11}|^2 - |\Delta|^2} \tag{4.85}$$

Output stability circle:

$$\text{Center} = \frac{(s_{22} - \Delta s_{11}^*)^*}{|s_{22}|^2 - |\Delta|^2} \qquad \text{Radius} = \frac{|s_{12}s_{21}|}{|s_{22}|^2 - |\Delta|^2} \qquad (4.86)$$

$$\Delta = s_{11}s_{22} - s_{12}s_{21}$$

s_{ij} = s-parameters of two-port analyzed for stability

* = complex conjugate

The stability circles separate stable from potentially unstable terminating impedance regions on the Smith Chart. Another way of looking at stability is to realize that terminating impedances in the unstable region cause the device to exhibit negative resistance at its terminals. Basic network theory says that if a two-port device exhibits negative resistance at one port, it must do so at the other port also. If the input stability circle intersects the Smith Chart, there must also be an output stability circle that cuts the Smith Chart in the output plane at the same frequency. The number of input and output stability circles intersecting the Smith Chart must be equal.

The corollary to this is that either input or output resistive loading can cure instability, and once an input stability circle moves outside the Smith Chart, there is no need to check the output; it will be stable also (and vice versa).

Operation in the unstable region is sometimes desirable for oscillators.

4.20 STATISTICAL ANALYSIS

It has become customary to specify performance parameters in terms of their statistical properties, to track or ensure high production yields. The most common way is to specify how many standard deviations the average is from the specification limit. For example, a 6-sigma design specifies that the average value is six standard deviations away from the specification limit. Other process capability indices are also used [22] to describe statistical properties of a process whose specification limits have been prescribed:

$$C_P = \frac{USL - LSL}{6\sigma} \qquad (4.87)$$

$$C_{PU} = \frac{USL - \mu}{3\sigma} \qquad (4.88)$$

$$C_{PL} = \frac{\mu - LSL}{3\sigma} \qquad (4.89)$$

$$C_{PK} = \text{Min}\{C_{PU}, C_{PL}\} \qquad (4.90)$$

C_P = process capability index

C_{PU} = upper capability index

C_{PL} = lower capability index

C_{PK} = process capability index adjusted for shift in mean value

USL = upper specification limit

LSL = lower specification limit

μ = average value of process

σ = standard deviation of process

Most specifications in RF work are single-ended, and we can use either C_{PU} or C_{PL} so that $C_\bullet$ simply becomes a measure of how far the average value is from the specification limit, expressed as a fraction of 3σ. The following example, shown in Figure 4.19, illustrates the concepts.

If the average measured sensitivity of a production run of receivers is 0.23 μV with a standard deviation of 0.07 μV, then we can calculate C_{PK} with respect to the specification limit of 0.35 μV as

$$C_{PK} = \frac{USL - \mu}{3\sigma} = \frac{0.35 - 0.23}{3(0.07)} = 0.571$$

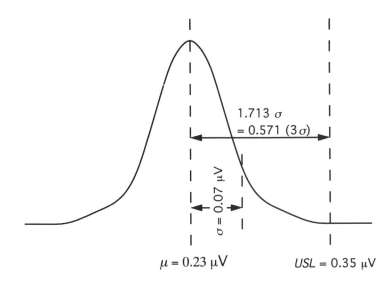

Figure 4.19 Difference between average performance μ, and specification limit *USL* expressed as a fraction of 3σ.

Such a low value of C_{PK} is typically not acceptable for large-volume production, because it would result in a failure rate of 4.3%. For $C_{PK} > 0.5$, the failure rate can be estimated from an approximation to the standard error function [23]:

$$\text{Failure rate} \approx \frac{1}{\sqrt{2\pi}} \exp[-4.5(C_{PK})^2] \left[\frac{1}{3C_{PK}} - \frac{1}{(3C_{PK})^3} + \frac{1.3}{(3C_{PK})^5} \right] \qquad (4.91)$$

Restriction: $C_{PK} > 0.5$

This type of statistical analysis demands much measured data, or a sophisticated computer prediction model must be available. Let us consider an example of the latter. Equation (1.12) in Subsection 1.1.3 on receiver-adjacent channel selectivity shows how selectivity is determined by five stage properties.

Each of these properties (LO SSB phase noise, synthesizer spurs, IF selectivity, IF bandwidth, and co-channel rejection) has a statistical distribution of values. To ascertain the impact of these statistical variations on overall selectivity, we can program the selectivity equation into a system simulator, such as Extend™ [24].

We can develop the required model, shown in Figure 4.20, assuming the following stage properties and their statistical distributions:

- IF selectivity uniformly distributed between 95 and 100 dB;
- IF bandwidth uniformly distributed between 11 and 13 kHz;
- Synthesizer spurious emissions uniformly distributed between 90 and 130 dB;
- VCO SSB phase noise Gaussian with mean −131 dBc/Hz and sigma 0.5 dB;
- Co-channel rejection Gaussian with mean 4.8 dB, sigma 0.6 dB.

The Eqn block contains the selectivity equation (1.12), the five random-number generators feed in the appropriate statistical distributions of the input parameters, and the whole simulation is run for thousands of trials, resulting in the displayed average and standard deviation parameters of overall selectivity, as well as a histogram of the expected selectivity values, shown in Figure 4.21.

The resulting selectivity has a mean of 84.3 dB with a standard deviation of 0.97 dB. We can achieve the specification limit of 80 dB with a C_{PK} of about 1.5, which really means that the specification limit is 1.5 times 3σ, or 4.5 standard deviations away from the average selectivity value. In this example, the failure rate for selectivity would be near 5 parts per million!

Statistical modeling is a powerful design and analysis tool and is not constrained to modeling just selectivity. Any performance parameter, which can be described by a mathematical combination of several inputs can be analyzed in this way. First, define the mathematical relationship, then feed in the randomly generated parameter values using the appropriate statistical distributions, and monitor the statistical properties of the calculated results. Ideally, the source of the parameter statistical distributions is Monte Carlo circuit analysis. In the example here, the IF

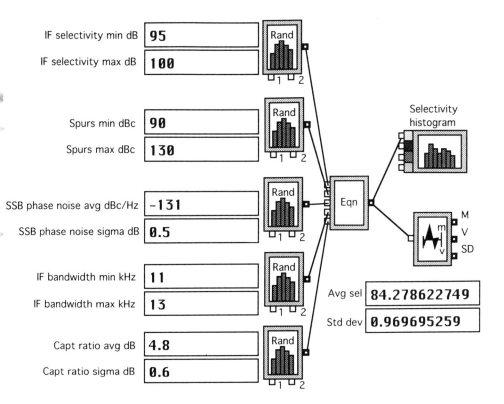

IF selectivity min dB	**95**	
IF selectivity max dB	**100**	
Spurs min dBc	**90**	
Spurs max dBc	**130**	
SSB phase noise avg dBc/Hz	**−131**	
SSB phase noise sigma dB	**0.5**	
IF bandwidth min kHz	**11**	
IF bandwidth max kHz	**13**	
Capt ratio avg dB	**4.8**	
Capt ratio sigma dB	**0.6**	

Selectivity histogram

Avg sel **84.278622749**

Std dev **0.969695259**

Figure 4.20 Statistical modeling of selectivity in Extend™.

section would first be analyzed using the appropriate component value distributions in a circuit simulator to come up with a statistical distribution for the IF bandwidth. This IF bandwidth distribution would then be used as an input to the system simulation. As you can see, this is just Monte Carlo analysis applied to systems, not just circuits.

It is important to keep in mind that C_{PK} calculation applies only to normal (Gaussian) distributions; it must be verified that the process under study is indeed Gaussian. While the central limit theorem states that the overall sum of independent random variables of any distribution has Gaussian distribution, we find in practice that many output parameters are strongly determined by one or two input parameters. A good example is noise figure of the first stage influencing receiver sensitivity. The statistical distribution of sensitivity will strongly correlate to first-stage noise figure, which may not be Gaussian.

The histogram in our example shows that selectivity is also not truly Gaussian but shows a longer "tail" toward lower values. The calculated C_{PK} number is, therefore, only an approximate measure of process capability in this case.

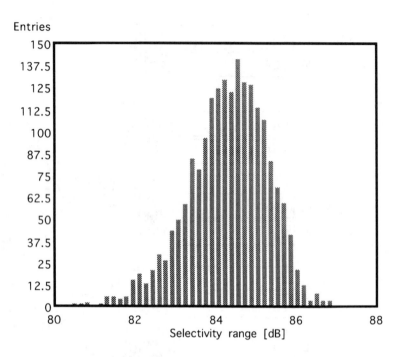

Figure 4.21 Histogram of simulated selectivity distribution.

Quantities expressed in decibels are rarely Gaussian. Always plot a histogram or perform skewness and kurtosis tests [25,26] to ensure that your parameter is Gaussian. If not, try its linear equivalent or transform the parameter into a form that is more Gaussian.

Computer tools can be used to perform statistical analysis on circuits or systems, using Monte Carlo algorithms. As of this writing (1994), the following computer programs can be used for statistical analysis and yield prediction. This list is not comprehensive by any means, and I apologize for any omissions.

MDS™ from Hewlett-Packard [27] (HP workstations). Yield analysis, nonlinear capability, linked to printed circuit layout.

Supercompact™ from Compact Software [28] (Workstations and PCs). Nonlinear and optoelectronic capability.

MMICAD™ from Optotek [29] (PCs). User-defined models.

RFDesigner™ from ingSOFT [30] (Macintosh). Statistical distributions can be assigned to device s-parameters.

Extend™ from Imagine That! [24] (Macintosh, PC). General-purpose system simulator with wide range of statistical distributions built in.

Gaussian From Uniform

A formula for generating numbers with a Gaussian distribution from a uniform distribution random-number generator is

$$x = \mu + \sigma\cos(2\pi RND)\sqrt{-2\ln(RND)} \qquad (4.92)$$

x = random real Gaussian variable of mean μ and standard deviation σ

μ = desired average of x

σ = desired standard deviation of x

RND = random real variable with uniform distribution between 0 and 1
   (a succession of numbers from a uniform distribution
   random-number generator).

Most computer programs and languages have a function that can generate RND.

4.21 TEMPERATURE

Three units of temperature are in common use today:

$$K = {}^\circ C + 273.15$$
$${}^\circ C = K - 273.15$$
$${}^\circ C = \frac{5}{9}({}^\circ F - 32)$$
$${}^\circ F = \frac{9}{5}({}^\circ C) + 32$$

4.22 TOLERANCE

The tolerances for most discrete components are part of the component's specification. For example, one can buy ±2% capacitors. There are many more types of components for which the tolerances are not usually known but must be estimated for statistical analysis and yield prediction. Table 4.1 provides an estimate of the plus and minus tolerances that you can expect from using some common components.

In addition to the value tolerance, you should also be concerned with tolerance introduced by the manufacturing process. Air-wound coils of low inductance value are particularly affected by the soldering process, as illustrated in Figure 4.22.

The coil on the left has larger inductance than the coil on the right, due to the soldering process and possibly to variation in strip length if the coil wire is insulated.

Table 4.1
Plus and Minus Tolerances of Common Parts

Bipolar transistor s-parameters	
$\|s_{11}\|$, $\|s_{22}\|$	10% to 20%
$\angle s_{11}$, $\angle s_{22}$	10° to 15°
$\|s_{21}\|$	10%
$\angle s_{21}$	15°
$\|s_{12}\|$	20%
$\angle s_{12}$	20°
Capacitor	
ESR	20%
Coaxial cable	
Z_0	5%
v_p	2%
α	10%
Inductor inductance	
Wire-wound coils	10% to 20%
With ferrite core	20% to 50%
Wound on ceramic core	5% to 10%
Inductor Q	20%
Printed spiral inductor	
Inductance uncertainty	1% to 5%
Repeatability	Better than 1%
Inductor Q	20%
Standard 1.5-mm-thick printed circuit board (G10, FR4) traces	
Z_0	8%
v_p	3%
α	20%
Capacitance	0.03 to 0.05 pF/mm²
Inductance	0.1 to 0.4 nH/mm depending on width
1 oz. copper trace thickness	0.036 mm ±5% unplated
	0.071 mm ±10% tin-plated
Printed directional coupler coupling	1 dB

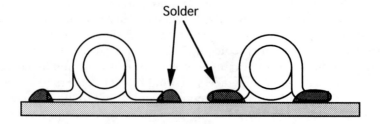

Figure 4.22 Inductance uncertainty introduced by soldering process.

4.23 TRANSMITTER POWER

The power output of a class-C amplifier is approximated by

$$P_0 \approx \frac{(V_{CC} - V_{sat})^2}{2R_L} \tag{4.93}$$

R_L = load resistance required at device output (Ω)
V_{CC} = dc supply voltage (V)
V_{sat} = device saturation voltage (V)
P_0 = desired RF power output (W)

Equation (4.93) illustrates the difficulty of designing broadband high-power amplifiers using a low dc voltage: High RF power requires a low value of R_L presented to the device. This means that the impedance transformation from this low resistance up to the system impedance becomes more extreme for higher powers, and the more extreme the impedance transformation, the narrower the operating bandwidth becomes (see Section 2.12).

4.24 UNITS AND CONSTANTS

dBm = unit of power, dB above 1 mW, independent of system impedance
0 dBm = 1 mW
dBmV = dB above 1 mV rms
V = rms voltage

Power and Voltage in a 50-Ω system:

$$dBm = 20 \log(\sqrt{20} \ V) \tag{4.94}$$
$$V = \sqrt{0.05} \ 10^{(dBm/20)} \tag{4.95}$$

Power and Voltage in a 75-Ω system:

$$dBm = 20 \log(\sqrt{13.33} \ V) \tag{4.96}$$
$$V = \sqrt{0.075} \ 10^{(dBm/20)} \tag{4.97}$$
$$dBmV = dBm + 48.75 \ dB \tag{4.98}$$
$$dBm = dBmV - 48.75 \ dB \tag{4.99}$$

Transmission line attenuation:

1 dB/m = 0.115 Nepers/m = 30.48 dB/100 ft
1 Neper/m = 8.686 dB/m = 264.7 dB/100 ft
α [dB/m] = α [Nepers/m] × 8.686

Noise:
0 Kelvin = −273.15 °C
Noise power at 300 K = −173.83 dBm/Hz (50 Ω system)
= −124.43 dBmV/Hz (75 Ω system)
≈ −59 dBmV in a TV channel bandwidth
Boltzmann's constant, k = 1.38062 × 10⁻²³ (J/K)
Conductivity of copper = 5.76 × 10⁷ siemens (mho)/m
Conductivity of silver = 6.17 × 10⁷ siemens/m
Conductivity of solder = 0.7 × 10⁷ siemens/m
Dielectric constant/loss tangent of Alumina = 9.7/0.0003
Dielectric constant/loss tangent of G-10, FR4 = 4.5/0.002
Dielectric constant/loss tangent of PTFE board = 2.2/0.0009
Dielectric constant/loss tangent of nylon = 3.9/0.04
Electron charge, e = −1.6021 × 10⁻¹⁹ C
Plane wave impedance of vacuum, η = 376.73 Ω
Permeability of vacuum, μ_0 = 4π × 10⁻⁷ H/m
Permittivity of vacuum, ϵ_0 = 8.854185 × 10⁻¹² F/m
Planck's constant, h = 6.6262 × 10⁻³⁴ J s
Speed of light in vacuum, c = 2.997925 × 10⁸ m/s

4.25 VELOCITY FACTOR

Velocity factor is the ratio between actual velocity of propagation and the speed of light in a vacuum. It determines the wavelength in the propagation medium.

$$v_p = \frac{v}{c} = \frac{1}{\sqrt{\epsilon_{\text{eff}}}} \qquad (4.100)$$

v_p = velocity factor, dimensionless, v_p < 1
v = group velocity of propagation in medium (m/s)
c = speed of light in free space, 2.9979 × 10⁸ (m/s)

$$\text{Effective dielectric constant } \epsilon_{\text{eff}} = \frac{1}{v_p^2} \qquad (4.101)$$

Approximate velocity factors for specific transmission line media are given in Table 4.2:

Table 4.2
Velocity Factors of Common Transmission Lines

Medium	$Z_0[\Omega]$	Width [mm]	v_p
G-10 (FR4)	50	2.7	0.528
microstrip	75	1.2	0.554
	100	0.54	0.553
Teflon/fiberglass	50	4.2	0.688
microstrip	75	2.1	0.702
	100	1.1	0.713
TV twin lead	300		0.82

4.26 VOLTAGE FROM FIELD STRENGTH

If the plane-wave electric field strength at a given antenna location is known, we can obtain the RF voltage at the antenna terminals from (4.5), provided that the plane wave and the antenna polarizations match.

$$V_r = 4.359 \times 10^6 \frac{E_r\sqrt{GR}}{f} \tag{4.102}$$

V_r = voltage at receiver antenna terminals, rms (V)

E_r = electric field strength of plane wave at antenna location (V/m)

G = gain of receiving antenna (linear)

R = receiver input impedance (Ω), matched to antenna

f = frequency (Hz)

4.27 WAVEGUIDE BEYOND CUTOFF

The formula for attenuation of waveguide beyond cutoff can be used to evaluate the effectiveness of shields, for coupling of energy inside nonresonant boxes, across gaps, and through small holes.

$$\alpha \approx \frac{2\pi}{\lambda_c}\sqrt{1 - \left(\frac{f}{f_c}\right)^2} \tag{4.103}$$

α = attenuation of waveguide below cutoff (Nepers/m)

$\quad$ (1 Neper/m = 8.686 dB/m = 2.647 dB/ft)

λ_c = cutoff wavelength (m)

f = frequency (Hz)

f_c = cutoff frequency (Hz) = v/λ_c

v = propagation velocity = $c/\sqrt{\epsilon_r}$ (m/s)

c = speed of light in free space, 2.9979×10^8 (m/s)

ϵ_r = dielectric constant of material filling the waveguide

Equation (4.103) holds for any shape of waveguide, with the following assumptions:

1. Cutoff frequency and cutoff wavelength are known for the waveguide shape. The cutoff wavelength for a long slot can be approximated by the following formula:

$$\lambda_c|_{TE10} \approx 2a$$

λ_c = cutoff wavelength for dominant TE_{10} mode (m)
a = longest transverse dimension of structure (m)
The cutoff wavelength for a circular waveguide (i.e., hole) is given by

$$\lambda_c|_{TE11} \approx 3.42r$$

λ_c = cutoff wavelength for circular waveguide dominant TE_{11} mode (m)
r = radius of hole (m)

2. The operating frequency is less than cutoff frequency, $f < f_c$.
3. Attenuation refers to the mode for which the f_c is appropriate, and the input of the waveguide is being excited in this mode, so that (4.103) gives the lowest possible attenuation. If the appropriate mode is not being excited, then the attenuation will be higher.
4. The receiving end of the waveguide is terminated in a nonresonant, perfectly absorbing load. If two circuits that are resonant at the same frequency are connected by a waveguide, the resulting attenuation cannot be predicted by (4.103) and in fact will be very low for high Q circuits.

To illustrate usage of (4.103), consider two circuits built on the same double-sided circuit board, each covered by its own surface shield, as shown in Figure 4.23. There is a row of through holes in the ground planes separating the two circuits.

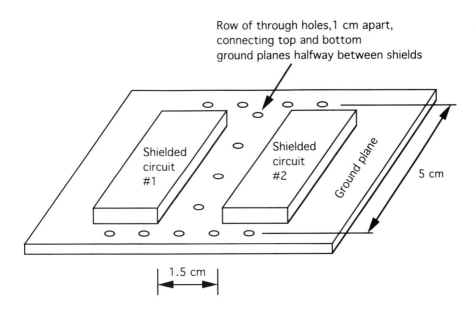

Row of through holes,1 cm apart, connecting top and bottom ground planes halfway between shields

Shielded circuit #1

Shielded circuit #2

Ground plane

5 cm

1.5 cm

Figure 4.23 Coupling between two shielded circuits through circuit board.

We would like to find the theoretical isolation between the two shielded compartments at 500 MHz, assuming that all the coupling happens across gaps between the through holes inside the circuit board and subject to the four assumptions listed above.

The approximate waveguide connection between the two compartments through the circuit board consists of three waveguides in series:

- *First waveguide:* 5-cm-wide waveguide from the edge of shield 1 to the row of holes. Cutoff wavelength = 2 × 0.05 = 0.1 m. Length of waveguide = 1.4/2 = 0.7 cm. Cutoff frequency assuming G-10 material (ϵ_r = 4.8) = 1.36 GHz.
- *Second waveguide:* 1-cm-wide waveguide representing the distance between the through holes. Cutoff wavelength = 2 × 0.01 = 0.02 m. Length of waveguide = 0.1 cm (diameter of through hole). Cutoff frequency = 6.82 GHz.
- *Third waveguide:* 5-cm-wide waveguide from row of holes to edge of shield 2. Cutoff wavelength = 2 × 0.05 = 0.1 m. Length of waveguide = 0.7 cm. Cutoff frequency = 1.36 GHz.

Figure 4.24 shows the equivalent circuit. The minimum theoretical attenuation at 500 MHz in the three waveguides is the sum of three terms derived from (4.103); each attenuation in Nepers per meter is multiplied by the length of the respective waveguide to arrive at the total attenuation:

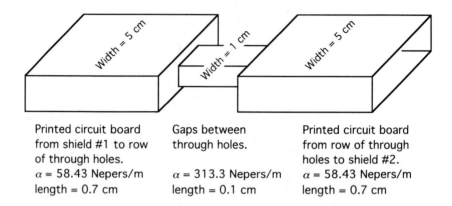

Printed circuit board from shield #1 to row of through holes.
α = 58.43 Nepers/m
length = 0.7 cm

Gaps between through holes.

α = 313.3 Nepers/m
length = 0.1 cm

Printed circuit board from row of through holes to shield #2.
α = 58.43 Nepers/m
length = 0.7 cm

Figure 4.24 Equivalent waveguide circuit between two shielded compartments.

$$\alpha = 58.43 \times 0.007 + 313.3 \times 0.001 + 58.43 \times 0.007 = 1.12 \text{ Nepers} \approx 10 \text{ dB.}$$

This is surprisingly low isolation between two apparently well-shielded circuits! We need more through holes between the two circuits for greater isolation. Keep in mind that the likelihood of exciting the correct TE_{10} mode between the two ground planes of the circuit board in one compartment and fully absorbing it in the other compartment is rather low, so the observed isolation will probably be higher than this theoretical minimum. However, if we have resonant circuits at both ends of the waveguide, then the attenuation through the waveguide will be very low and will be primarily determined by the Qs of the resonant circuits. Also note that to a first approximation, the circuit board thickness does not enter into the calculations, shielding effectiveness is not greatly improved by use of thinner gaps, and as long as a gap or a seam exists, no matter how thin it is, energy will tend to leak through it.

In this example we assumed that the correct waveguide mode was being excited and also fully absorbed at the various circuit-waveguide interfaces. The more general analysis of a plane wave incident on a conducting shield with openings shows that reflection of the incident energy by the metal shield accounts for most of the shielding effectiveness [31], with waveguide attenuation contributing additional attenuation only if the shield is relatively thick compared to the size of the openings.

REFERENCES

[1] Lerner, R. G., ed., *Encyclopedia of Physics,* 2d ed. New York: VCH Publishers, 1991.
[2] Carr, J. J., *Elements of Microwave Electronics Technology.* San Diego: Harcourt Brace Jovanovic, 1989, p. 381.
[3] Erst, S. J., *Electronics Equations Handbook.* Blue Ridge Summit, PA: Tab Books, 1989, p. 174.
[4] American Radio Relay League, *The ARRL Handbook for Radio Amateurs,* Newington, CT, 1992, pp. 2–17.

[5] Bowick, C., *RF Circuit Design*, 1st ed. Carmel, IN: SAMS Books, 1982.

[6] ingSOFT Ltd, "Modeling Real Inductors," *RFDesigner Newsletter*, Vol.2, No. 1, Willowdale, ONT, March 1989, p. 6.

[7] Lang, D., "Broadband Model Predicts s-Parameters of Spiral Inductors," *Microwaves & RF*, January 1988, p. 107.

[8] Wadell, B. C., *Transmission Line Design Handbook*. Norwood, MA: Artech House, 1991, pp. 389–404.

[9] Zverev, A. I., *Handbook of Filter Synthesis*. New York: John Wiley and Sons, 1967, p. 530.

[10] Robins, W. P., *Phase Noise in Signal Sources*. London: Peter Peregrinus Ltd., 1982, p. 53.

[11] Borras, J. A., "Improve Synthesized Transceiver Performance and Reliability by Simple Screening of the VCO Active Device," *Proceedings of RF Technology Expo 86*, 1986, Anaheim, CA, p. 87.

[12] Rhea, R. W., *Oscillator Design and Computer Simulation*. Englewood Cliffs, NJ: Prentice-Hall, 1990, p. 32.

[13] Vizmuller, P., "A Simple Frequency Divider Circuit," *RF Design*, September, 1986, p. 76.

[14] Zverev, A. I., *Handbook of Filter Synthesis*. New York: John Wiley and Sons, 1967, p. 501.

[15] Vizmuller, P., *Filters With Helical and Folded Helical Resonators*, Dedham, MA: Artech House, 1987, p. 61.

[16] Carr, J. J., *Elements of Microwave Electronics Technology*. San Diego: Harcourt Brace Jovanovic, 1989, p. 12.

[17] Wadell, B. C., *Transmission Line Design Handbook*. Norwood, MA: Artech House, 1991, p. 25.

[18] Kraus, J. D., and K. R. Carver, *Electromagnetics*, 2d ed. New York: McGraw-Hill, 1973, p. 404.

[19] GraphSoft, Inc., MiniCAD+, 8370 Court Ave, Suite B100, Ellicott City, MD 21043, (410) 290-5114.

[20] Ramo, S., J. R. Whinnery, and T. Van Duzer, *Fields and Waves in Communication Electronics*. New York: John Wiley & Sons, 1965, p. 36.

[21] Yip, P. C. L., *High-Frequency Circuit Design and Measurements*. London: Chapman & Hall, 1990, p. 115.

[22] Sullivan, L. P., "Reducing Variability: A New Approach to Quality," *Quality Progress*, July 1984.

[23] Gellert, W., ed., *The VNR Concise Encyclopedia of Mathematics*, 1st American ed. New York: Van Nostrand Reinhold Co., 1975, p. 625.

[24] Imagine That, Inc., *Extend Software*, 6830 Via Del Oro, Suite 230, San Jose, CA 95119, (408) 365-0305.

[25] Kurtz, M., *Handbook of Applied Mathematics for Engineers and Scientists*. New York: McGraw-Hill, 1991, p. 8.28.

[26] Meehan, M. D., *Yield and Reliability in Microwave Circuit and System Design*, Norwood, MA: Artech House, 1993, p. 193.

[27] Hewlett-Packard, *HP Microwave and RF Design Systems*, Santa Rosa Systems Division, 1400 Fountaingrove Parkway, Santa Rosa, CA 95403.

[28] *Supercompact CAE Software*, Compact Software, Inc., 201 McLean Boulevard, Paterson, NJ 07504, (201) 881-1200.

[29] Optotek Ltd., *MMICAD*, 62 Steacie Dr., Kanata, ONT K2K2A9, Canada, (613) 591-0336

[30] ingSOFT Limited, *RFDesigner Software*, 213 Dunview Ave., Willowdale, ONT M2N 4H9, Canada, (416) 730-9611.

[31] Interference Control Technologies, Inc., *EMC Considerations for the Design of Communication Transceivers*, Seminar notes, Gainsville, VA, 1994, p. 3.29.

Workbook Software Installation and User's Guide

The workbook software included with this book contains equations of the preceding chapters encoded in spreadsheet format, organized as Microsoft® Excel™ workbooks. The workbooks are arranged in the same sequence as the book's chapters and sections, making a particular calculation easy to find.

The workbook software can work with both IBM compatible and Apple® Macintosh® computers, and uses the familiar Excel user interface for data entry, calculation, and printing.

A.1 INSTALLATION AND USAGE ON IBM AND COMPATIBLE COMPUTERS

Required items:

1. Computer capable of running Windows 3.1 or later.
2. Microsoft Excel version 4.0 or later.
3. 4 MB or more memory.
4. 3-1/2″ high density floppy disk drive.
5. VGA Color monitor recommended.

Run Excel, and open the desired workbook. You may want to copy the workbooks to your hard disk for faster opening and closing of documents.

A.2 INSTALLATION AND USAGE ON MACINTOSH COMPUTERS USING SYSTEM 7.5 OR LATER

Required items:

1. Computer with 4 MB or more memory and hard disk drive.
2. Microsoft Excel version 4.0 or later. Set its preferred size to at least 2 MB.
3. 3-1/2" high density superdrive.
4. Color monitor recommended.
5. Make sure the "PC Exchange" control panel is ON.

Insert diskette into drive; it should be recognized as a "PC" diskette. Launch Excel and open the required workbooks from within Excel. Save workbooks on your hard disk for faster opening and saving of documents.

A.3 INSTALLATION AND USAGE ON MACINTOSH COMPUTERS USING SYSTEM SOFTWARE PRIOR TO SYSTEM 7.5

Required items:

1. Computer with 4 MB or more memory and hard disk drive.
2. Microsoft Excel version 4.0 or later. Set its preferred size to at least 2 MB.
3. 3-1/2" high density superdrive.
4. Apple File Exchange software (Included with your original system software).
5. Color monitor recommended.

Procedure:

a. Launch Apple File Exchange and insert floppy disk into drive.
b. Transfer all the files from diskette to your hard disk using the Default Translation.
c. Quit Apple File Exchange and launch Microsoft Excel; allocate 2 MB or more memory to Excel. It will open the Workbooks much faster with more memory.
d. Open each workbook from within Excel and re-save to give it the proper icon.

A.4 USING THE WORKBOOKS

When you first open a Workbook, it displays a table of contents listing. Simply double-click on the section of interest to access the relevant calculation. It is very important to keep in mind that information in most worksheets takes up more than one screen; always scroll down to check for additional information further down on the screen. Each Workbook also contains a Help section as the first entry in the table of contents for quick reference.

All worksheets follow the same conventions: Cells formatted in bold character style are available for user input; the active cell can be advanced by pressing the TAB key. Cells formatted in red are the calculated outputs, italics are intermediate calculations, and blue cells are used for annotating diagrams with linked values.

The worksheets have been protected against inadvertent changes. If you wish to make changes, select "Unprotect Document . . ." under the Options menu. The individual documents can also be unbound from the Workbook file; consult the Excel manual for more information on manipulating individual documents in Workbooks.

The ReadMe file, which is also included on the disk, contains additional details on software installation, usage, and history.

This software is provided for your convenience; when you become familiar with its organization, feel free to group your favorite calculations into your own Workbooks: Create a new Workbook and drag items from one of the existing Workbooks' table of contents into your new Workbook's table of contents.

A.5 DISCLAIMER OF WARRANTY ON SOFTWARE

This software is provided "as is" without warranty of any kind. Neither Artech House nor the author warrant, guarantee or make any representation regarding the use or the results of use of the software in terms of correctness, accuracy, reliability, currentness, or otherwise. The entire risk as to the results and performance of the software is assumed by the user. In no event shall Artech House or the author be liable for any direct, indirect, consequential or incidental damages arising out of the use or inability to use the software.

Glossary

1/2 IF	Half–IF spurious response frequency. This receiver spurious is $0.5 \times f_{IF}$ below the receive frequency for low-side injection and $0.5 \times f_{IF}$ above the receive frequency for high-side injection.
ACPR	Adjacent channel protection ratio.
AGC	Automatic gain control.
AM	Amplitude modulation, measured in percent.
ASK	Amplitude shift keying.
Balun	Balanced to unbalanced transformer.
BER	Bit error rate.
CATV	Cable TV.
Capture ratio	S/N at detector input required for desired demodulated signal performance (usually audio S/N or BER).
Compression point	Usually 1-dB compression point, the level at which device gain decreases by 1 dB due to limiting, or generation of harmonics.
C_{pk}	Process capability index; a statistical measure of how far the process average is away from the specification limit(s).
C_{pl}	Process capability index with respect to the lower specification limit.
C_{pu}	Process capability index with respect to the upper specification limit.
CW	Continuous wave; refers to an unmodulated sine–wave signal.
dBm	Power level relative to 1 mW.
dBmV	Voltage level relative to 1 mV rms.
DPSK	Differential phase shift keying.
DQPSK	Differential quadrature phase shift keying.
DS	Direct–sequence technique for spread–spectrum communication.

Duplex	Mode of operation when both transmitter and receiver are on at the same time.
DUT	Device under test.
ENR	Excess noise ratio of a noise source.
erfc	Complementary error function.
ESR	Equivalent series resistance of a capacitor.
FET	Field effect transistor.
FH	Frequency–hopping technique for spread-spectrum communication.
FM	Frequency modulation, measured in Hz peak frequency deviation.
FSK	Frequency shift keying.
High-side injection	First LO frequency is higher than RF signal.
Hybrid	Commonly used for 90° power splitter or any signal splitting or combining network.
IC	Integrated circuit.
IF	Intermediate frequency, mixer output.
IM	Intermodulation distortion, usually third order.
Image	Most common receiver spurious response frequency. It is $2 \times f_{IF}$ below the receive frequency for low-side injection and $2 \times f_{IF}$ above the receive frequency for high-side injection.
Injection	Refers to the high-level signal fed into a mixer, local oscillator. High-side injection means that LO frequency is above RF frequency and vice versa.
Intercept point	Fictitious power level at which desired signal and distortion products are equal in amplitude.
IP2	Second-order input intercept point.
IP3	Third-order input intercept point.
IPn	nth-order input intercept point.
JFET	Junction field effect transistor.
k	Stability factor, coupling coefficient.
K	Absolute temperature in Kelvins.
K_v	VCO tuning sensitivity in Hz/V.
$\mathscr{L}$	SSB phase noise.
LC	Inductor/capacitor; refers to circuits that use lumped inductor and capacitor components (as opposed to transmission lines or acoustic components).
LO	Local oscillator, high-level input into mixer.
Low-side injection	First LO frequency is lower than RF signal.
MMIC	Microwave monolithic integrated circuit.

OOK	On–off keying, ASK with 100% modulation.
PA	Power amplifier.
PIN	Refers to a special diode optimized with respect to transit time for low RF resistance (p–type doping, intrinsic, n–type doping three-layer junction structure).
PLL	Phase–locked loop.
PM	Phase modulation.
Prestage gain	Refers to the sum of gains in decibels or the product of linear gains up to but excluding the stage itself.
PSK	Phase shift keying.
Q	Quality factor; a measure of stored energy/dissipated energy, also a measure of bandwidth.
QPSK	Quadrature phase shift keying.
RF	Radio frequency; also low-level input to mixer.
RND	Software function for generating random numbers between 0 and 1 with uniform distribution.
RSSI	Received signal strength indicator.
RX	Receiver.
s	$j\omega$, complex frequency.
S	Receiver sensitivity in dBm.
SAW	Surface acoustic wave device (high Q but poor frequency stability).
SBN	SSB phase noise.
Schottky diode	Metal–semiconductor majority carrier diode.
Second image	Spurious response frequency in dual conversion receivers, separated from the desired frequency by twice the second IF frequency.
Simplex	Refers to a communication mode where the transmitter and the receiver are never on at the same time; transmit and receive operation is sequential.
SINAD	Ratio of signal to noise and distortion; 12-dB SINAD is the usual measure of receiver sensitivity in North America for narrowband communication voice systems.
S/N or *SNR*	Signal-to-noise ratio.
SSB	Single–sideband AM modulated signal.
SSB phase noise	Refers to single–sideband PM noise relatively close to the carrier, usually at an offset equal to the system channel spacing.
Steering line	Control voltage responsible for VCO frequency tuning.
TE	Transverse electric mode in a waveguide.

TEM	Transverse electromagnetic mode in a transmission medium. This mode supports dc transmission. Coaxial cable is an example of TEM transmission medium.
T/N	Ratio of tone to noise; 20-dB T/N is the usual measure of receiver sensitivity in Europe for narrowband communication voice systems.
Transducer gain	Ratio of power absorbed by load to power available from source.
TRL	Transmission line (as identified in Figure 2.101).
TX	Transmitter.
VCO	Voltage-controlled oscillator.
VSWR	Voltage standing wave ratio.
Wideband noise	Refers to single–sideband AM or PM noise relatively far away from the carrier, usually at an offset greater than 10 times the system channel spacing.
α	Attenuation of transmission line.
Γ	Reflection coefficient.
δ	Skin depth.
ϵ	Permittivity.
ϵ_{eff}	Effective dielectric constant.
ϵ_r	Dielectric constant, relative permittivity.
θ	Thermal resistance.
ζ	Damping ratio.
λ	Wavelength.
μ	Permeability, or average value.
σ	Conductivity, or standard deviation.
τ	Group delay, time duration.
ω	Radian frequency.

About the Author

Peter Vizmuller is an 18-year veteran of the RF communications industry, designing high performance radio frequency circuits and systems for Motorola. His latest assignment was as a product architect of a major new communication product involving over 100 staff-years of effort by engineers in four countries. He holds three patents, and has published several articles as well as a book on helical resonator filters. He is a two-time winner of the annual design contest sponsored by *RFDesign* magazine.

Mr. Vizmuller graduated from the University of Toronto in 1977, with a subsequent Master's degree in engineering in 1981. He contributed his expertise in system and circuit modeling, statistical techniques and optimization algorithms to the development of several leading CAEE software products.

He was born in Slovakia in 1954, raised and educated in Canada, and now lives in Richmond Hill, Ontario, Canada with his wife and three children.

Index

The Artech House Microwave Library

Microwave Transmission Line Couplers, J. A. G. Malherbe

Microwave Tubes, A. S. Gilmour, Jr.

Microwaves: Industrial, Scientific, and Medical Applications, J. Thuery

Microwaves Made Simple: Principles and Applicatons, Stephen W. Cheung, Frederick H. Levien *et al.*

MMIC Design: GaAs FETs and HEMTs, Peter H. Ladbrooke

Modern GaAs Processing Techniques, Ralph Williams

Modern Microwave Measurements and Techniques, Thomas S. Laverghetta

Monolithic Microwave Integrated Circuits: Technology and Design, Ravender Goyal *et al.*

Nonuniform Line Microstrip Directional Couplers, Sener Uysal

PC Filter: Electronic Filter Design Software and User's Guide, Michael G. Ellis, Sr.

PLL: Linear Phase-Locked Loop Control Systems Analysis Software and User's Manual, Eric L. Unruh

RF Design Guide: Systems, Circuits, and Equations, Peter Vizmuller

Scattering Parameters of Microwave Networks with Multiconductor Transmission Lines: Software & User's Manual, A. R. Djordjevic *et al.*

Solid-State Microwave Power Oscillator Design, Eric Holzman and Ralston Robertson

Terrestrial Digital Microwave Communications, Ferdo Ivanek *et al.*

Time-Domain Response of Multiconductor Transmission Lines: Software and User's Manual, A. R. Djordjevic *et al.*

Transmission Line Design Handbook, Brian C. Waddell

Yield and Reliability in Microwave Circuit and System Design, Michael Meehan and John Purviance

For further information on these and other Artech House titles, contact:

Artech House
685 Canton Street
Norwood, MA 02062
617-769-9750
Fax: 617-769-6334
Telex: 951-659
email: artech@world.std.com

Artech House
Portland House, Stag Place
London SW1E 5XA England
+44 (0) 171-973-8077
Fax: +44 (0) 171-630-0166
Telex: 951-659
bookco@artech.demon.co.uk